Student Solutions Manual for Zill's
A FIRST COURSE IN DIFFERENTIAL EQUATIONS
with Modeling Applications
Sixth Edition

WARREN S. WRIGHT
Loyola Marymount University
and
CAROL D. WRIGHT

PWS Publishing Company

Brooks/Cole Publishing Company

I(T)P® An International Thomson Publishing Company

Pacific Grove • Albany • Bonn • Boston • Cincinnati • Detroit • London • Madrid • Melbourne
Mexico City • New York • Paris • San Francisco • Singapore • Tokyo • Toronto • Washington

Sponsoring Editor: *Gary Ostedt*
Assistant Editor: *Beth Wilbur*
Marketing Team: *Patrick Farrant and Deborah Petit*
Editorial Assistant: *Nancy Conti*
Production Coordinator: *Mary Vezilich*
Cover Design: *Jennifer Mackres*
Printing and Binding: *Malloy Lithographing*

For more information, contact:

BROOKS/COLE PUBLISHING COMPANY
511 Forest Lodge Road
Pacific Grove, CA 93950
USA

International Thomson Editores
Campos Eliseos 385, Piso 7
Col. Polanco
11560 México D. F. México

International Thomson Publishing Europe
Berkshire House 168-173
High Holborn
London WC1V 7AA
England

International Thomson Publishing GmbH
Königswinterer Strasse 418
53227 Bonn
Germany

Thomas Nelson Australia
102 Dodds Street
South Melbourne, 3205
Victoria, Australia

International Thomson Publishing Asia
221 Henderson Road
#05-10 Henderson Building
Singapore 0315

Nelson Canada
1120 Birchmount Road
Scarborough, Ontario
Canada M1K 5G4

International Thomson Publishing Japan
Hirakawacho Kyowa Building, 3F
2-2-1 Hirakawacho
Chiyoda-ku, Tokyo 102
Japan

Printed in the United States of America

5   4   3   2   1

ISBN 0-534-95578-9

Table of Contents

1 Introduction to Differential Equations

3. First-order; nonlinear because of yy'.

6. Second-order; nonlinear because of $\sin y$.

9. Third-order; linear.

12. From $y = 8$ we obtain $y' = 0$, so that $y' + 4y = 0 + 4(8) = 32$.

15. From $y = 5 \tan 5x$ we obtain $y' = 25 \sec^2 5x$. Then

$$y' = 25 \sec^2 5x = 25 \left(1 + \tan^2 5x\right) = 25 + (5 \tan 5x)^2 = 25 + y^2.$$

18. First write the differential equation in the form $2xy + \left(x^2 + 2y\right) y' = 0$. Implicitly differentiating $x^2 y + y^2 = c_1$ we obtain $2xy + \left(x^2 + 2y\right) y' = 0$.

21. Implicitly differentiating $y^2 = c_1 \left(x + \frac{1}{4}c_1\right)$ we obtain $y' = c_1/2y$. Then

$$2xy' + y(y')^2 = \frac{c_1 x}{y} + \frac{c_1^2}{4y} = \frac{y^2}{y} = y.$$

24. Differentiating $P = ac_1 e^{at} / \left(1 + bc_1 e^{at}\right)$ we obtain

$$\frac{dP}{dt} = \frac{\left(1 + bc_1 e^{at}\right) a^2 c_1 e^{at} - ac_1 e^{at} \cdot abc_1 e^{at}}{\left(1 + bc_1 e^{at}\right)^2}$$

$$= \frac{ac_1 e^{at}}{1 + bc_1 e^{at}} \cdot \frac{\left[a \left(1 + bc_1 e^{at}\right) - abc_1 e^{at}\right]}{1 + bc_1 e^{at}} = P(a - bP).$$

27. First write the differential equation in the form $y' = \dfrac{-x^2 - y^2}{x^2 - xy}$. Then $c_1(x + y)^2 = xe^{y/x}$ implies

$c_1 = \dfrac{xe^{y/x}}{(x + y)^2}$ and implicit differentiation gives $2c_1(x + y)(1 + y') = xe^{y/x}\dfrac{xy' - y}{x^2} + e^{y/x}$. Solving for y' we obtain

$$y' = \frac{e^{y/x} - \frac{y}{x}e^{y/x} - 2c_1(x + y)}{2c_1(x + y) - e^{y/x}} = \frac{1 - \frac{y}{x} - \frac{2x}{x + y}}{\frac{2x}{x + y} - 1} = \frac{-x^2 - y^2}{x^2 - xy}.$$

30. From $y = e^{2x} + xe^{2x}$ we obtain $\dfrac{dy}{dx} = 3e^{2x} + 2xe^{3x}$ and $\dfrac{d^2y}{dx^2} = 8e^{2x} + 4xe^{2x}$ so that $\dfrac{d^2y}{dx^2} - 4\dfrac{dy}{dx} + 4y = 0$.

33. From $y = \ln|x + c_1| + c_2$ we obtain $y' = \dfrac{1}{x + c_1}$ and $y'' = \dfrac{-1}{(x + c_1)^2}$, so that $y'' + (y')^2 = 0$.

36. From $y = x\cos(\ln x)$ we obtain $y' = -\sin(\ln x) + \cos(\ln x)$ and $y'' = \dfrac{-1}{x}\cos(\ln x) - \dfrac{1}{x}\sin(\ln x)$, so that $x^2 y'' - xy' + 2y = 0$.

39. From $y = x^2 e^x$ we obtain $y' = x^2 e^x + 2xe^x$, $y'' = x^2 e^x + 4xe^x 2e^x$, and $y''' = x^2 e^x + 6xe^x + 6e^x$, so that $y''' - 3y'' + 3y' - y = 0$.

42. From $y = \begin{cases} 0, & x < 0 \\ x^3, & x \geq 0 \end{cases}$ we obtain $y' = \begin{cases} 0, & x < 0 \\ 3x^2, & x \geq 0 \end{cases}$ so that $(y')^2 = \begin{cases} 0, & x < 0 \\ 9x^4, & x \geq 0. \end{cases}$

45. From $y = e^{mx}$ we obtain $y' = me^{mx}$ and $y'' = m^2 e^{mx}$. Then $y'' - 5y' + 6y = 0$ implies

$$m^2 e^{mx} - 5me^{mx} + 6e^{mx} = (m - 2)(m - 3)e^{mx} = 0.$$

Since $e^{mx} > 0$ for all x, $m = 2$ and $m = 3$. Thus $y = e^{2x}$ and $y = e^{3x}$ are solutions.

48. Using $y' = mx^{m-1}$ and $y'' = m(m-1)x^{m-2}$ and substituting into the differential equation we obtain $x^2 y'' + 6xy' + 4y = [m(m-1) + 6m + 4]x^m$. The right side will be zero provided m satisfies

$$m(m - 1) + 6m + 4 = m^2 + 5m + 4 = (m + 4)(m + 1) = 0.$$

Thus, $m = -4, -1$ and two solutions of the differential equation on the interval $0 < x < \infty$ are $y = x^{-4}$ and $y = x^{-1}$.

Exercises 1.2

3. For $f(x, y) = \dfrac{y}{x}$ we have $\dfrac{\partial f}{\partial y} = \dfrac{1}{x}$. Thus the differential equation will have a unique solution in any region where $x \neq 0$.

6. For $f(x, y) = \dfrac{x^2}{1 + y^3}$ we have $\dfrac{\partial f}{\partial y} = \dfrac{-3x^2 y^2}{(1 + y^3)^2}$. Thus the differential equation will have a unique solution in any region where $y \neq -1$.

9. For $f(x, y) = x^3 \cos y$ we have $\dfrac{\partial f}{\partial y} = -x^2 \sin y$. Thus the differential equation will have a unique solution in the entire plane.

12. Two solutions are $y = 0$ and $y = x^2$. (Also, any constant multiple of x^2 is a solution.)

15. We identify $f(x, y) = \sqrt{y^2 - 9}$ and $\partial f / \partial y = y^2 / \sqrt{y^2 - 9}$. Since $\partial f / \partial y$ is discontinuous for $|y| < 3$, the differential equation is not guaranteed to have a unique solution at $(2, -3)$.

18. (a) Since $1+y^2$ and its partial derivative with respect to y are continuous everywhere in the plane, the differential equation has a unique solution through every point in the plane.

(b) Since $\dfrac{d}{dx}(\tan x) = \sec^2 x = 1 + \tan^2 x$ and $\tan 0 = 0$, $y = \tan x$ satisfies the differential equation and the initial condition. Since $-2 < \pi/2 < 2$ and $\tan x$ is undefined for $x = \pi/2$, $y = \tan x$ is not a solution on the interval $-2 < x < 2$.

(c) Since $\tan x$ is continuous and differentiable on $(-\pi/2, \pi/2)$ and not defined at the endpoints of this interval, this is the largest interval of validity for which $y = \tan x$ is a solution of $y' = 1+y^2$, $y(0) = 0$.

21. Setting $x = 0$ and $y = -1/3$ we have $1/(1 + c_1) = -1/3$ so $c_1 = -4$. A solution of the initial-value problem is $y = 1/(1 - 4e^{-x})$.

24. From the initial conditions we obtain the system

$$c_1 c + c_2 c^{-1} = 0$$

$$c_1 e - c_2 e^{-1} = e.$$

Solving we get $c_1 = \frac{1}{2}$ and $c_2 = -\frac{1}{2}e^2$. A solution of the initial-value problem is $y = \frac{1}{2}e^x - \frac{1}{2}e^{2-x}$.

Exercises 1.3

3. The differential equation is $x'(t) = r - kx(t)$ where $k > 0$.

6. The rate at which salt is entering the tank is

$$R_1 = (3\text{ gal/min}) \cdot (2\text{ lb/gal}) = 6\text{ lb/min}.$$

Since the solution is pumped out at a slower rate, it is accumulating at the rate of $(3-2)\text{gal/min} = 1\text{ gal/min}$. After t minutes there are $300 + t$ gallons of brine in the tank. The rate at which salt is leaving is

$$R_2 = (2\text{ gal/min}) \cdot \left(\frac{A}{300 + t}\text{ lb/gal}\right) = \frac{2A}{300 + t}\text{ lb/min}.$$

The differential equation is

$$\frac{dA}{dt} = 6 - \frac{2A}{300 + t}.$$

9. Since $i = \dfrac{dq}{dt}$ and $L\dfrac{d^2q}{dt^2} + R\dfrac{dq}{dt} = E(t)$ we obtain $L\dfrac{di}{dt} + Ri = E(t)$.

12. The differential equation is $\dfrac{dA}{dt} = k_1(M - A) - k_2 A$.

15. The net force acting on the mass is

$$F = ma = m\frac{d^2x}{dt^2} = -k(s+x) + mg = -kx + mg - ks.$$

Since the condition of equilibrium is $mg = ks$, the differential equation is

$$m\frac{d^2x}{dt^2} = -ks.$$

Chapter 1 Review Exercises

3. False; since $y = 0$ is a solution.

6. Third-order; ordinary; nonlinear because of $\sin xy$.

9. From $y = x + \tan x$ we obtain $y' = 1 + \sec^2 x$, and $y'' = 2\sec^2 x \tan x$. Using $1 + \tan^2 x = \sec^2 x$ we have $y' + 2xy = 2 + x^2 + y^2$.

12. From $y = \sin 2x + \cosh 2x$ we obtain $y^{(4)} = 16\sin 2x + 16\cosh 2x$ so that $y^{(4)} - 16y = 0$.

15. $y = \frac{1}{2}x^2$

18. $y = \sqrt{x}$

21. For all values of y, $y^2 - 2y \geq -1$. Avoiding left– and right–hand derivatives, we then must have $x^2 - x - 1 > -1$. That is, $x < 0$ or $x > 1$.

24. From Newton's second law we obtain $m\dfrac{dv}{dt} = \dfrac{1}{2}mg - \mu\dfrac{\sqrt{3}}{2}mg$ or $\dfrac{dv}{dt} = 16\left(1 - \sqrt{3}\,\mu\right)$.

2 First-Order Differential Equations

_____ **Exercises 2.1** _____

In many of the following problems we will encounter an expression of the form $\ln|g(y)| = f(x) + c$. To solve for $g(y)$ we exponentiate both sides of the equation. This yields $|g(y)| = e^{f(x)+c} = e^c e^{f(x)}$ which implies $g(y) = \pm e^c e^{f(x)}$. Letting $c_1 = \pm e^c$ we obtain $g(y) = c_1 e^{f(x)}$.

3. From $dy = -e^{-3x}\,dx$ we obtain $y = \dfrac{1}{3}e^{-3x} + c$.

6. From $dy = 2xe^{-x}dx$ we obtain $y = -2xe^{-x} - 2e^{-x} + c$.

9. From $\dfrac{1}{y^3}\,dy = \dfrac{1}{x^2}\,dx$ we obtain $y^{-2} = \dfrac{2}{x} + c$.

12. From $\left(\dfrac{1}{y} + 2y\right)\,dy = \sin x\,dx$ we obtain $\ln|y| + y^2 = -\cos x + c$.

15. From $\dfrac{y}{2+y^2}\,dy = \dfrac{x}{4+x^2}\,dx$ we obtain $\ln|2+y^2| = \ln|4+x^2| + c$ or $2+y^2 = c_1\left(4+x^2\right)$.

18. From $\dfrac{y^2}{y+1}\,dy = \dfrac{1}{x^2}\,dx$ we obtain $\dfrac{1}{2}y^2 - y + \ln|y+1| = -\dfrac{1}{x} + c$ or $\dfrac{1}{2}y^2 - y + \ln|y+1| = -\dfrac{1}{x} + c_1$.

21. From $\dfrac{1}{S}\,dS = k\,dr$ we obtain $S = ce^{kr}$.

24. From $\dfrac{1}{N}\,dN = \left(te^{t+2} - 1\right)\,dt$ we obtain $\ln|N| = te^{t+2} - e^{t+2} - t + c$.

27. From $\dfrac{e^{2y} - y}{e^y}\,dy = -\dfrac{\sin 2x}{\cos x}\,dx = -\dfrac{2\sin x \cos x}{\cos x}\,dx$ or $\left(e^y - ye^{-y}\right)\,dy = -2\sin x\,dx$ we obtain

$e^y + ye^{-y} + e^{-y} = 2\cos x + c.$

30. From $\dfrac{y}{(1+y^2)^{1/2}}\,dy = \dfrac{x}{(1+x^2)^{1/2}}\,dx$ we obtain $\left(1+y^2\right)^{1/2} = \left(1+x^2\right)^{1/2} + c$.

33. From $\dfrac{y-2}{y+3}\,dy = \dfrac{x-1}{x+4}\,dx$ or $\left(1 - \dfrac{5}{y-3}\right)\,dy = \left(1 - \dfrac{5}{x+4}\right)\,dx$ we obtain

$$y - 5\ln|y-3| = x - 5\ln|x+4| + c \quad\text{or}\quad \left(\dfrac{x+4}{y-3}\right)^5 = c_1 e^{x-y}.$$

36. From $\sec y\,\dfrac{dy}{dx} + \sin x \cos y - \cos x \sin y = \sin x \cos y + \cos x \sin y$ we find $\sec y\,dy = 2\sin y \cos x\,dx$ or

$\dfrac{1}{2\sin y \cos y}\,dy = \csc 2y\,dy = \cos x\,dx$. Then $\dfrac{1}{2}\ln|\csc 2y - \cot 2y| = \sin x + c$.

5

39. From $\dfrac{1}{y^2}\,dy = \dfrac{1}{e^x + e^{-x}}\,dx = \dfrac{e^x}{(e^x)^2 + 1}\,dx$ we obtain $-\dfrac{1}{y} = \tan^{-1} e^x + c.$

42. From $\dfrac{1}{1 + (2y)^2}\,dy = \dfrac{-x}{1 + (x^2)^2}\,dx$ we obtain

$$\frac{1}{2}\tan^{-1} 2y = -\frac{1}{2}\tan^{-1} x^2 + c \quad \text{or} \quad \tan^{-1} 2y + \tan^{-1} x^2 = c_1.$$

Using $y(1) = 0$ we find $c_1 = \pi/4$. The solution of the initial-value problem is

$$\tan^{-1} 2y + \tan^{-1} x^2 = \frac{\pi}{4}.$$

45. From $\dfrac{1}{x^2 + 1}\,dx = 4\,dy$ we obtain $\tan^{-1} x = 4y + c$. Using $x(\pi/4) = 1$ we find $c = -3\pi/4$. The solution of the initial-value problem is $\tan^{-1} x = 4y - \dfrac{3\pi}{4}$ or $x = \tan\left(4y - \dfrac{3\pi}{4}\right)$.

48. From $\dfrac{1}{1 - 2y}\,dy = dx$ we obtain $-\dfrac{1}{2}\ln|1 - 2y| = x + c$ or $1 - 2y = c_1 e^{-2x}$. Using $y(0) = 5/2$ we find $c_1 = -4$. The solution of the initial-value problem is $1 - 2y = -4e^{-2x}$ or $y = 2e^{-2x} + \dfrac{1}{2}$.

51. By inspection a singular solution is $y = 1$.

54. Separating variables we obtain $\dfrac{dy}{(y - 1)^2} = dx$. Then $-\dfrac{1}{y - 1} = x + c$ and $y = \dfrac{x + c - 1}{x + c}$. Setting $x = 0$ and $y = 1.01$ we obtain $c = -100$. The solution is $y = \dfrac{x - 101}{x - 100}$.

Exercises 2.2

3. Let $M = 5x + 4y$ and $N = 4x - 8y^3$ so that $M_y = 4 = N_x$. From $f_x = 5x + 4y$ we obtain $f = \dfrac{5}{2}x^2 + 4xy + h(y)$, $h'(y) = -8y^3$, and $h(y) = -2y^4$. The solution is $\dfrac{5}{2}x^2 + 4xy - 2y^4 = c.$

6. Let $M = 4x^3 - 3y\sin 3x - y/x^2$ and $N = 2y - 1/x + \cos 3x$ so that $M_y = -3\sin 3x - 1/x^2$ and $N_x = 1/x^2 - 3\sin 3x$. The equation is not exact.

9. Let $M = y^3 - y^2\sin x - x$ and $N = 3xy^2 + 2y\cos x$ so that $M_y = 3y^2 - 2y\sin x = N_x$. From $f_x = y^3 - y^2\sin x - x$ we obtain $f = xy^3 + y^2\cos x - \dfrac{1}{2}x^2 + h(y)$, $h'(y) = 0$, and $h(y) = 0$. The solution is $xy^3 + y^2\cos x - \dfrac{1}{2}x^2 = c.$

12. Let $M = 2x/y$ and $N = -x^2/y^2$ so that $M_y = -2x/y^2 = N_x$. From $f_x = 2x/y$ we obtain $f = \dfrac{x^2}{y} + h(y)$, $h'(y) = 0$, and $h(y) = 0$. The solution is $x^2 = cy.$

15. Let $M = 1 - 3/x + y$ and $N = 1 - 3/y + x$ so that $M_y = 1 = N_x$. From $f_x = 1 - 3/x + y$ we obtain

$f = x - 3\ln|x| + xy + h(y)$, $h'(y) = 1 - \dfrac{3}{y}$, and $h(y) = y - 3\ln|y|$. The solution is

$x + y + xy - 3\ln|xy| = c$

18. Let $M = -2y$ and $N = 5y - 2x$ so that $M_y = -2 = N_x$. From $f_x = -2y$ we obtain $f = -2xy + h(y)$,

$h'(y) = 5y$, and $h(y) = \dfrac{5}{2}y^2$. The solution is $-2xy + \dfrac{5}{2}y^2 = c$.

21. Let $M = 4x^3 + 4xy$ and $N = 2x^2 + 2y - 1$ so that $M_y = 4x = N_x$. From $f_x = 4x^3 + 4xy$ we obtain
$f = x^4 + 2x^2 y + h(y)$, $h'(y) = 2y - 1$, and $h(y) = y^2 - y$. The solution is $x^4 + 2x^2 y + y^2 - y = c$.

24. Let $M = 1/x + 1/x^2 - y/\left(x^2 + y^2\right)$ and $N = ye^y + x/\left(x^2 + y^2\right)$ so that

$M_y = \left(y^2 - x^2\right)/\left(x^2 + y^2\right)^2 = N_x$. From $f_x = 1/x + 1/x^2 - y/\left(x^2 + y^2\right)$ we obtain

$f = \ln|x| - \dfrac{1}{x} - \arctan\left(\dfrac{x}{y}\right) + h(y)$, $h'(y) = ye^y$, and $h(y) = ye^y - e^y$. The solution is

$$\ln|x| - \dfrac{1}{x} - \arctan\left(\dfrac{x}{y}\right) + ye^y - e^y = c.$$

27. Let $M = 4y + 2x - 5$ and $N = 6y + 4x - 1$ so that $M_y = 4 = N_x$. From $f_x = 4y + 2x - 5$ we
obtain $f = 4xy + x^2 - 5x + h(y)$, $h'(y) = 6y - 1$, and $h(y) - 3y^2 - y$. The general solution is
$4xy + x^2 - 5x + 3y^2 - y = c$. If $y(-1) = 2$ then $c = 8$ and the solution of the initial-value problem
is $4xy + x^2 - 5x + 3y^2 - y = 8$.

30. Let $M = y^2 + y\sin x$ and $N = 2xy - \cos x - 1/\left(1 + y^2\right)$ so that $M_y = 2y + \sin x = N_x$. From

$f_x = y^2 + y\sin x$ we obtain $f = xy^2 - y\cos x + h(y)$, $h'(y) = \dfrac{-1}{1 + y^2}$, and $h(y) = -\tan^{-1} y$. The

general solution is $xy^2 - y\cos x - \tan^{-1} y = c$. If $y(0) = 1$ then $c = -1 - \pi/4$ and the solution of
the initial-value problem is $xy^2 - y\cos x - \tan^{-1} y = -1 - \dfrac{\pi}{4}$.

33. Equating $M_y = 4xy + e^x$ and $N_x = 4xy + ke^x$ we obtain $k = 1$.

36. Since $f_x = M(x, y) = y^{1/2} x^{-1/2} + x\left(x^2 + y\right)^{-1}$ we obtain $f = 2y^{1/2} x^{1/2} + \dfrac{1}{2}\ln\left|x^2 + y\right| + h(x)$ so that

$f_y = y^{-1/2} x^{1/2} + \dfrac{1}{2}\left(x^2 + y\right)^{-1} + h'(x)$. Let $N(x, y) = y^{-1/2} x^{1/2} + \dfrac{1}{2}\left(x^2 + y\right)^{-1}$.

39. Let $M = -x^2 y^2 \sin x + 2xy^2 \cos x$ and $N = 2x^2 y \cos x$ so that $M_y = -2x^2 y \sin x + 4xy\cos x = N_x$.
From $f_y = 2x^2 y\cos x$ we obtain $f = x^2 y^2 \cos x + h(y)$, $h'(y) = 0$, and $h(y) = 0$. The solution of the
differential equation is $x^2 y^2 \cos x = c$.

42. Let $M = \left(x^2 + 2xy - y^2\right)/\left(x^2 + 2xy + y^2\right)$ and $N = \left(y^2 + 2xy - x^2\right)/\left(y^2 + 2xy + x^2\right)$ so that
$M_y = -4xy/(x + y)^3 = N_x$. From $f_x = \left(x^2 + 2xy + y^2 - 2y^2\right)/(x + y)^2$ we obtain

$f = x + \dfrac{2y^2}{x+y} + h(y)$, $h'(y) = -1$, and $h(y) = -y$. The solution of the differential equation is $x^2 + y^2 = c(x+y)$.

Exercises 2.3

3. For $y' + 4y = \dfrac{4}{3}$ an integrating factor is $e^{\int 4dx} = e^{4x}$ so that $\dfrac{d}{dx}\left[e^{4x}y\right] = \dfrac{4}{3}e^{4x}$ and $y = \dfrac{1}{3} + ce^{-4x}$ for $-\infty < x < \infty$.

6. For $y' - y = e^x$ an integrating factor is $e^{-\int dx} = e^{-x}$ so that $\dfrac{d}{dx}\left[e^{-x}y\right] = 1$ and $y = xe^x + ce^x$ for $-\infty < x < \infty$.

9. For $y' + \dfrac{1}{x}y = \dfrac{1}{x^2}$ an integrating factor is $e^{\int(1/x)dx} = x$ so that $\dfrac{d}{dx}[xy] = \dfrac{1}{x}$ and $y = \dfrac{1}{x}\ln x + \dfrac{c}{x}$ for $0 < x < \infty$.

12. For $\dfrac{dx}{dy} - x = y$ an integrating factor is $e^{-\int dy} = e^{-y}$ so that $\dfrac{d}{dy}\left[e^{-y}x\right] = ye^{-y}$ and $x = -y-1+ce^y$ for $-\infty < y < \infty$.

15. For $y' + \dfrac{e^x}{1+e^x}y = 0$ an integrating factor is $e^{\int[e^x/(1+e^x)]dx} = 1 + e^x$ so that $\dfrac{d}{dx}\left[1 + e^x y\right] = 0$ and $y = \dfrac{c}{1+e^x}$ for $-\infty < x < \infty$.

18. For $y' + (\cot x)y = 2\cos x$ an integrating factor is $e^{\int \cot x\, dx} = \sin x$ so that $\dfrac{d}{dx}\left[(\sin x)\,y\right] = 2\sin x \cos x$ and $y = \sin x + c\csc x$ for $0 < x < \pi$.

21. For $y' + \left(1 + \dfrac{2}{x}\right)y = \dfrac{e^x}{x^2}$ an integrating factor is $e^{\int[1+(2/x)]dx} = x^2 e^x$ so that $\dfrac{d}{dx}\left[x^2 e^x y\right] = e^{2x}$ and $y = \dfrac{1}{2}\dfrac{e^x}{x^2} + \dfrac{ce^{-x}}{x^2}$ for $0 < x < \infty$.

24. For $y' + \dfrac{2\sin x}{(1 - \cos x)}y = \tan x(1 - \cos x)$ an integrating factor is $e^{\int[2\sin x/(1-\cos x)]dx} = (1 - \cos x)^2$ so that $\dfrac{d}{dx}\left[(1 - \cos x)^2 y\right] = \tan x - \sin x$ and $y(1 - \cos x)^2 = \ln|\sec x| + \cos x + c$ for $0 < x < \pi/2$.

27. For $y' + \left(3 + \dfrac{1}{x}\right)y = \dfrac{e^{-3x}}{x}$ an integrating factor is $e^{\int[3+(1/x)]dx} = xe^{3x}$ so that $\dfrac{d}{dx}\left[xe^{3x}y\right] = 1$ and $y = e^{-3x} + \dfrac{ce^{-3x}}{x}$ for $0 < x < \infty$.

30. For $y' + \dfrac{2}{x}y = \dfrac{1}{x}(e^x + \ln x)$ an integrating factor is $e^{\int(2/x)dx} = x^2$ so that $\dfrac{d}{dx}\left[x^2 y\right] = xe^x + x\ln x$ and $x^2 y = xe^x - e^x + \dfrac{x^2}{2}\ln x - \dfrac{1}{4}x^2 + c$ for $0 < x < \infty$.

33. For $\dfrac{dx}{dy} + \left(2y + \dfrac{1}{y}\right)x = 2$ an integrating factor is $e^{\int [2y+(1/y)]dy} = ye^{y^2}$ so that $\dfrac{d}{dy}\left[ye^{y^2}x\right] = 2ye^{y^2}$ and $x = \dfrac{1}{y} + \dfrac{1}{y}ce^{-y^2}$ for $0 < y < \infty$.

36. For $\dfrac{dP}{dt} + (2t - 1)P = 4t - 2$ an integrating factor is $e^{\int(2t-1)\,dt} = e^{t^2-t}$ so that

$\dfrac{d}{dt}\left[Pe^{t^2-t}\right] = (4t - 2)e^{t^2-t}$ and $P = 2 + ce^{t-t^2}$ for $-\infty < t < \infty$.

39. For $y' + (\cosh x)y = 10\cosh x$ an integrating factor is $e^{\int \cosh x\,dx} = e^{\sinh x}$ so that

$\dfrac{d}{dx}\left[e^{\sinh x}y\right] = 10(\cosh x)e^{\sinh x}$ and $y = 10 + ce^{-\sinh x}$ for $-\infty < x < \infty$.

42. For $y' - 2y = x\left(e^{3x} - e^{2x}\right)$ an integrating factor is $e^{-\int 2dx} = e^{-2x}$ so that $\dfrac{d}{dx}\left[e^{-2x}y\right] = xe^{x} - x$ and $y = xe^{3x} - e^{3x} - \frac{1}{2}x^2e^{2x} + ce^{2x}$ for $-\infty < x < \infty$. If $y(0) = 2$ then $c = 3$ and

$y = xe^{3x} - e^{3x} - \dfrac{1}{2}x^2e^{2x} + 3e^{2x}$.

45. For $y' + (\tan x)y = \cos^2 x$ an integrating factor is $e^{\int \tan x\,dx} = \sec x$ so that $\dfrac{d}{dx}\left[(\sec x)\,y\right] = \cos x$ and $y = \sin x \cos x + c\cos x$ for $-\pi/2 < x < \pi/2$. If $y(0) = -1$ then $c = -1$ and $y = \sin x \cos x - \cos x$.

48. For $y' + \left(1 + \dfrac{2}{x}\right)y = \dfrac{2}{x}e^{-x}$ an integrating factor is $e^{\int(1+2/x)dx} = x^2e^{x}$ so that $\dfrac{d}{dx}\left[x^2e^{x}y\right] = 2x$ and $y = e^{-x} + \dfrac{c}{x^2}e^{-x}$ for $0 < x < \infty$. If $y(1) = 0$ then $c = -1$ and $y = e^{-x} - \dfrac{1}{x^2}e^{-x}$.

51. For $y' + 2y = f(x)$ an integrating factor is e^{2x} so that

$$ye^{2x} = \begin{cases} \frac{1}{2}e^{2x} + c_1, & 0 \le x \le 3; \\ c_2, & x > 3. \end{cases}$$

If $y(0) = 0$ then $c_1 = -1/2$ and for continuity we must have $c_2 = \frac{1}{2}e^6 - \frac{1}{2}$ so that

$$y = \begin{cases} \frac{1}{2}\left(1 - e^{-2x}\right), & 0 \le x \le 3; \\ \frac{1}{2}\left(e^6 - 1\right)e^{-2x}, & x > 3. \end{cases}$$

54. For

$$y' + \dfrac{2x}{1 + x^2}y = \begin{cases} \dfrac{x}{1+x^2}, & 0 \le x < 1; \\ \dfrac{-x}{1+x^2}, & x \ge 1 \end{cases}$$

an integrating factor is $1 + x^2$ so that

$$\left(1 + x^2\right) y = \begin{cases} \frac{1}{2}x^2 + c_1, & 0 \le x < 1; \\ -\frac{1}{2}x^2 + c_2, & x \ge 1. \end{cases}$$

If $y(0) = 0$ then $c_1 = 0$ and for continuity we must have $c_2 = 1$ so that

$$y = \begin{cases} \frac{1}{2} - \frac{1}{2(1+x^2)}, & 0 \le x < 1; \\ \frac{3}{2(1+x^2)} - \frac{1}{2}, & x \ge 1. \end{cases}$$

57. An integrating factor for $y' - 2xy = 1$ is e^{-x^2}. Thus

$$\frac{d}{dx}[e^{-x^2}y] = e^{-x^2}$$

$$e^{-x^2}y = \int_0^x e^{-t^2} \, dt = \mathrm{erf}(x) + c$$

and

$$y = e^{x^2}\mathrm{erf}(x) + ce^{x^2}.$$

From $y(1) = 1$ we get $1 = e\,\mathrm{erf}(1) + ce$, so that $c = e^{-1} - \mathrm{erf}(1)$. Thus

$$y = e^{x^2}\mathrm{erf}(x) + (e^{-1} - \mathrm{erf}(1))e^{x^2}$$

$$= e^{x^2-1} + e^{x^2}(\mathrm{erf}(x) - \mathrm{erf}(1)).$$

Exercises 2.4

3. Letting $x = vy$ we have

$$vy(v\,dy + y\,dv) + (y - 2vy)\,dy = 0$$

$$vy\,dv + \left(v^2 - 2v + 1\right)dy = 0$$

$$\frac{v\,dv}{(v-1)^2} + \frac{dy}{y} = 0$$

$$\ln|v - 1| - \frac{1}{v-1} + \ln|y| = c$$

$$\ln\left|\frac{x}{y} - 1\right| - \frac{1}{x/y - 1} + \ln y = c$$

$$(x - y)\ln|x - y| - y = c(x - y).$$

6. Letting $y = ux$ we have

$$\left(u^2 x^2 + ux^2\right) dx + x^2 (u\,dx + x\,du) = 0$$

$$\left(u^2 + 2u\right) dx + x\,du = 0$$

$$\frac{dx}{x} + \frac{du}{u(u+2)} = 0$$

$$\ln|x| + \frac{1}{2}\ln|u| - \frac{1}{2}\ln|u+2| = c$$

$$\frac{x^2 u}{u+2} = c_1$$

$$x^2\frac{y}{x} = c_1\left(\frac{y}{x} + 2\right)$$

$$x^2 y = c_1(y + 2x).$$

9. Letting $y = ux$ we have

$$-ux\,dx + (x + \sqrt{u}\,x)(u\,dx + x\,du) = 0$$

$$(x + x\sqrt{u})\,du + u^{3/2}\,dx = 0$$

$$\left(u^{-3/2} + \frac{1}{u}\right)du + \frac{dx}{x} = 0$$

$$-2u^{-1/2} + \ln|u| + \ln|x| = c$$

$$\ln|y/x| + \ln|x| = 2\sqrt{x/y} + c$$

$$y(\ln|y| - c)^2 = 4x.$$

12. Letting $y = ux$ we have

$$\left(x^2 + 2u^2 x^2\right) dx - ux^2(u\,dx + x\,du) = 0$$

$$\left(1 + u^2\right) dx - ux\,du = 0$$

$$\frac{dx}{x} - \frac{u\,du}{1+u^2} = 0$$

$$\ln|x| - \frac{1}{2}\ln\left(1 + u^2\right) = c$$

$$\frac{x^2}{1+u^2} = c_1$$

$$x^4 = c_1\left(y^2 + x^2\right).$$

11

Exercises 2.4

Using $y(-1) = 1$ we find $c_1 = 1/2$. The solution of the initial-value problem is $2x^4 = y^2 + x^2$.

15. From $y' + \dfrac{1}{x}y = \dfrac{1}{x}y^{-2}$ and $w = y^3$ we obtain $\dfrac{dw}{dx} + \dfrac{3}{x}w = \dfrac{3}{x}$. An integrating factor is x^3 so that $x^3 w = x^3 + c$ or $y^3 = 1 + cx^{-3}$.

18. From $y' - \left(1 + \dfrac{1}{x}\right)y = y^2$ and $w = y^{-1}$ we obtain $\dfrac{dw}{dx} + \left(1 + \dfrac{1}{x}\right)w = -1$. An integrating factor is xe^x so that $xe^x w = -xe^x + e^x + c$ or $y^{-1} = -1 + \dfrac{1}{x} + \dfrac{c}{x}e^{-x}$.

21. From $y' - \dfrac{2}{x}y = \dfrac{3}{x^2}y^4$ and $w = y^{-3}$ we obtain $\dfrac{dw}{dx} + \dfrac{6}{x}w = -\dfrac{9}{x^2}$. An integrating factor is x^6 so that $x^6 w = -\dfrac{9}{5}x^5 + c$ or $y^{-3} = -\dfrac{9}{5}x^{-1} + cx^{-6}$. If $y(1) = \dfrac{1}{2}$ then $c = \dfrac{49}{5}$ and $y^{-3} = -\dfrac{9}{5}x^{-1} + \dfrac{49}{5}x^{-6}$.

24. Let $u = x + y$ so that $du/dx = 1 + dy/dx$. Then $\dfrac{du}{dx} - 1 = \dfrac{1 - u}{u}$ or $u\,du = dx$. Thus $\dfrac{1}{2}u^2 = x + c$ or $u^2 = 2x + c_1$, and $(x + y)^2 = 2x + c_1$.

27. Let $u = y - 2x + 3$ so that $du/dx = dy/dx - 2$. Then $\dfrac{du}{dx} + 2 = 2 + \sqrt{u}$ or $\dfrac{1}{\sqrt{u}}\,du = dx$. Thus $2\sqrt{u} = x + c$ and $2\sqrt{y - 2x + 3} = x + c$.

30. Let $u = 3x + 2y$ so that $du/dx = 3 + 2\,dy/dx$. Then $\dfrac{du}{dx} = 3 + \dfrac{2u}{u + 2} = \dfrac{5u + 6}{u + 2}$ and $\dfrac{u + 2}{5u + 6}\,du = dx$.
Now
$$\frac{u + 2}{5u + 6} = \frac{1}{5} + \frac{4}{25u + 30}$$
so we have
$$\int\left(\frac{1}{5} + \frac{4}{25u + 30}\right)du = dx$$
and $\dfrac{1}{5}u + \dfrac{4}{25}\ln|25u + 30| = x + c$. Thus
$$\frac{1}{5}(3x + 2y) + \frac{4}{25}\ln|75x + 50y + 30| = x + c.$$
Setting $x = -1$ and $y = -1$ we obtain $c = \dfrac{4}{5}\ln 95$. The solution is
$$\frac{1}{5}(3x + 2y) + \frac{4}{25}\ln|75x + 50y + 30| = x + \frac{4}{5}\ln 95$$
or
$$5y - 5x + 2\ln|75x + 50y + 30| = 10\ln 95.$$

_____ **Chapter 2 Review Exercises** _____

3. separable, exact, linear in x and y

6. separable, linear in x, Bernoulli

9. Bernoulli

12. exact, linear in y

15. Separating variables we obtain

$$\cos^2 x \, dx = \frac{y}{y^2 + 1} \, dy \implies \frac{1}{2}x + \frac{1}{4}\sin 2x = \frac{1}{2}\ln\left(y^2 + 1\right) + c$$

$$\implies 2x + \sin 2x = 2\ln\left(y^2 + 1\right) + c.$$

18. Write the differential equation in the form $(3y^2 + 2x)dx + (4y^2 + 6xy)dy = 0$. Letting $M = 3y^2 + 2x$ and $N = 4y^2 + 6xy$ we see that $M_y = 6y - N_x$ so the differential equation is exact. From $f_x = 3y^2 + 2x$ we obtain $f = 3xy^2 + x^2 + h(y)$. Then $f_y = 6xy + h'(y) = 4y^2 + 6xy$ and $h'(y) = 4y^2$ so $h(y) = \frac{4}{3}y^3$. The general solution is

$$3xy^2 + x^2 + \frac{4}{3}y^3 = c.$$

21. Separating variables we obtain

$$y \ln y \, dy = te^t dt \implies \frac{1}{2}y^2 \ln|y| - \frac{1}{4}y^2 = te^t - e^t + c.$$

If $y(1) = 1$, $c = -1/4$. The solution is $2y^2 \ln|y| - y^2 = 4te^t - 4e^t - 1$.

24. The differential equation is Bernoulli. Using $w = y^{-1}$ we obtain $-xy^2 \dfrac{dw}{dx} + 4y = x^4 y^2$ or

$$\frac{dw}{dx} - \frac{4}{x}w = -x^3. \text{ An integrating factor is } x^{-4}, \text{ so}$$

$$\frac{d}{dx}\left[x^{-4}w\right] = -\frac{1}{x} \implies x^{-4}w = -\ln x + c$$

$$\implies w = -x^4 \ln x + cx^4$$

$$\implies y = \left(cx^4 - x^4 \ln x\right)^{-1}.$$

If $y(1) = 1$ then $c = 1$ and $y = \left(x^4 - x^4 \ln x\right)^{-1}$.

3 Modeling with First-Order Differential Equations

―――――― **Exercises 3.1** ――――――

3. Let $P = P(t)$ be the population at time t. From $dP/dt = kt$ and $P(0) = P_0 = 500$ we obtain $P = 500e^{kt}$. Using $P(10) = 575$ we find $k = \frac{1}{10}\ln 1.15$. Then $P(30) = 500e^{3\ln 1.15} \approx 760$ years.

6. Let $N = N(t)$ be the amount at time t. From $dN/dt = kt$ and $N(0) = 100$ we obtain $N = 100e^{kt}$. Using $N(6) = 97$ we find $k = \frac{1}{6}\ln 0.97$. Then $N(24) = 100e^{(1/6)(\ln 0.97)24} = 100(0.97)^4 \approx 88.5$ mg.

9. Let $I = I(t)$ be the intensity, t the thickness, and $I(0) = I_0$. If $dI/dt = kI$ and $I(3) = 0.25I_0$ then $I = I_0e^{kt}$, $k = \frac{1}{3}\ln 0.25$, and $I(15) = 0.00098I_0$.

12. Assume that $dT/dt = k(T-5)$ so that $T = 5+ce^{kt}$. If $T(1) = 55°$ and $T(5) = 30°$ then $k = -\frac{1}{4}\ln 2$ and $c = 59.4611$ so that $T(0) = 64.4611°$.

15. Assume $L\,di/dt + Ri = E(t)$, $L = 0.1$, $R = 50$, and $E(t) = 50$ so that $i = \frac{3}{5} + ce^{-500t}$. If $i(0) = 0$ then $c = -3/5$ and $\lim_{t\to\infty} i(t) = 3/5$.

18. Assume $R\,dq/dt + (1/c)q = E(t)$, $R = 1000$, $C = 5 \times 10^{-6}$, and $E(t) = 200$. Then $q = \frac{1}{1000} + ce^{-200t}$ and $i = -200ce^{-200t}$. If $i(0) = 0.4$ then $c = -\frac{1}{500}$, $q(0.005) = 0.003$ coulombs, and $i(0.005) = 0.1472$ amps. As $t \to \infty$ we have $q \to \frac{1}{1000}$.

21. From $dA/dt = 4 - A/50$ we obtain $A = 200 + ce^{-t/50}$. If $A(0) = 30$ then $c = -170$ and $A = 200 - 170e^{-t/50}$.

24. From $\dfrac{dA}{dt} = 10 - \dfrac{10A}{500 - (10-5)t} = 10 - \dfrac{2A}{100-t}$ we obtain $A = 1000 - 10t + c(100-t)^2$. If $A(0) = 0$ then $c = -\dfrac{1}{10}$. The tank is empty in 100 minutes.

27. **(a)** From $m\,dv/dt = mg - kv$ we obtain $v = gm/k + ce^{-kt/m}$. If $v(0) = v_0$ then $c = v_0 - gm/k$ and the solution of the initial-value problem is

$$v = \frac{gm}{k} + \left(v_0 - \frac{gm}{k}\right)e^{-kt/m}.$$

(b) As $t \to \infty$ the limiting velocity is gm/k.

(c) From $ds/dt = v$ and $s(0) = s_0$ we obtain

$$s = \frac{gm}{k}t - \frac{m}{k}\left(v_0 - \frac{gm}{k}\right)e^{-kt/m} + s_0 + \frac{m}{k}\left(v_0 - \frac{gm}{k}\right).$$

30. Separating variables we obtain

$$\frac{dP}{P} = k\cos t\, dt \implies \ln|P| = k\sin t + c \implies P = c_1 e^{k\sin t}.$$

If $P(0) = P_0$ then $c_1 = P_0$ and $P = P_0 e^{k\sin t}$.

33. (a) Letting $t = 0$ correspond to 1790 we have $P(0) = 3.929$ and $P(t) = 3.929 e^{kt}$. Using $t = 10$, which corresponds to 1800, we have

$$5.308 = P(10) = 3.929 e^{10k}.$$

This implies that $k = 0.03$, so that

$$P(t) = 3.929 e^{0.03t}.$$

(b)

Year	Census Population	Predicted Population	Error	% Error
1790	3.929	3.929	0.000	0.00
1800	5.308	5.308	0.000	0.00
1810	7.240	7.171	0.069	0.95
1820	9.638	9.688	-0.050	-0.52
1830	12.866	13.088	-0.222	-1.73
1840	17.069	17.682	-0.613	-3.59
1850	23.192	23.888	-0.696	-3.00
1860	31.433	32.272	-0.839	-2.67
1870	38.558	43.599	-5.041	-13.07
1880	50.156	58.901	-8.745	-17.44
1890	62.948	79.574	-16.626	-26.41
1900	75.996	107.503	-31.507	-41.46
1910	91.972	145.234	-53.262	-57.91
1920	105.711	196.208	-90.497	-85.61
1930	122.775	265.074	-142.299	-115.90
1940	131.669	358.109	-226.440	-171.98
1950	150.697	483.798	-333.101	-221.04

Exercises 3.2

3. From $\dfrac{dP}{dt} = P\left(10^{-1} - 10^{-7}P\right)$ and $P(0) = 5000$ we obtain $P = \dfrac{500}{0.0005 + 0.0995 e^{-0.1t}}$ so that $P \to 1{,}000{,}000$ as $t \to \infty$. If $P(t) = 500{,}000$ then $t = 52.9$ months.

6. From Problem 5 we have $P = e^{a/b} e^{-ce^{-bt}}$ so that

$$\frac{dP}{dt} = bce^{a/b - bt} e^{-ce^{-bt}} \quad \text{and} \quad \frac{d^2P}{dt^2} = b^2 c e^{a/b - bt} e^{-ce^{-bt}}\left(ce^{-bt} - 1\right).$$

Setting $d^2P/dt^2 = 0$ and using $c = a/b - \ln P_0$ we obtain $t = (1/b)\ln(a/b - \ln P_0)$ and $P = e^{a/b-1}$.

9. If $\alpha \neq \beta$, $\dfrac{dX}{dt} = k(\alpha - X)(\beta - X)$, and $X(0) = 0$ then $\left(\dfrac{1/(\beta - \alpha)}{\alpha - X} + \dfrac{1/(\alpha - \beta)}{\beta - X}\right) dX = k\, dt$ so that

$$X = \frac{\alpha\beta - \alpha\beta e^{(\alpha-\beta)kt}}{\beta - \alpha e^{(\alpha-\beta)kt}}. \text{ If } \alpha = \beta \text{ then } \frac{1}{(\alpha - X)^2}\,dX = k\,dt \quad \text{and} \quad X = \alpha - \frac{1}{kt + c}.$$

12. In this case the differential equation is

$$\frac{dh}{dt} = -\frac{0.6}{25}\sqrt{h} = -\frac{3}{125}\sqrt{h}.$$

Separating variables and integrating we have $2h^{1/2} = -\frac{3}{125}t + c_1$ or $h^{1/2} = -\frac{3}{250}t + c_2$. From $h(0) = 20$ we find $c_2 = \sqrt{20}$, so $h = \left(\sqrt{20} - \frac{3}{250}t\right)^2$. Solving $h(t) = 0$ for t we find that the tank empties in $\frac{250}{3}\sqrt{20}$ s ≈ 6.2 min.

15. (a) Separating variables we obtain

$$\frac{m\,dv}{mg - kv^2} = dt$$

$$\frac{1}{g}\frac{dv}{1 - (kv/mg)^2} = dt$$

$$\frac{\sqrt{mg}}{\sqrt{k}\,g}\frac{\sqrt{k/mg}\,dv}{1 - (\sqrt{k}\,v/\sqrt{mg})^2} = dt$$

$$\sqrt{\frac{m}{kg}}\tanh^{-1}\frac{\sqrt{k}\,v}{\sqrt{mg}} = t + c$$

$$\tanh^{-1}\frac{\sqrt{k}\,v}{\sqrt{mg}} = \sqrt{\frac{kg}{m}}\,t + c_1.$$

Thus the velocity at time t is

$$v(t) = \sqrt{\frac{mg}{k}}\tanh\left(\sqrt{\frac{kg}{m}}\,t + c_1\right).$$

Setting $t = 0$ and $v = v_0$ we find $c_1 = \tanh^{-1}(\sqrt{k}\,v_0/\sqrt{mg})$.

(b) Since $\tanh t \to 1$ as $t \to \infty$, we have $v \to \sqrt{mg/k}$ as $t \to \infty$.

(c) Integrating the expression for $v(t)$ in part (a) we obtain

$$s(t) = \sqrt{\frac{mg}{k}}\int \tanh\left(\sqrt{\frac{kg}{m}}\,t + c_1\right)dt = \frac{m}{k}\ln\left[\cosh\left(\sqrt{\frac{kg}{m}}\,t + c_1\right)\right] + c_2.$$

Setting $t = 0$ and $s = s_0$ we find $c_2 = s_0 - \ln\cosh c_1$.

16

18. (a) We have $dP/dt = P(a - bP)$ with $P(0) = 3.929$ million. Using separation of variables we obtain

$$P(t) = \frac{3.929a}{3.929b + (a - 3.929b)e^{-at}} = \frac{a/b}{1 + (a/3.929b - 1)e^{-at}}$$

$$= \frac{c}{1 + (c/3.929 - 1)e^{-at}} .$$

At $t = 60(1850)$ the population is 23.192 million, so

$$23.192 = \frac{c}{1 + (c/3.929 - 1)e^{-60a}}$$

or $c = 23.192 + 23.192(c/3.929 - 1)e^{-60a}$. At $t = 120(1910)$

$$91.972 = \frac{c}{1 + (c/3.929 - 1)e^{-120a}}$$

or $c = 91.972 + 91.972(c/3.929 - 1)(e^{-60a})^2$. Combining the two equations for c we get

$$\left(\frac{(c - 23.192)/23.192}{c/3.929 - 1}\right)^2 \left(\frac{c}{3.929} - 1\right) = \frac{c - 91.972}{91.972}$$

or

$$91.972(3.929)(c - 23.192)^2 = (23.192)^2(c - 91.972)(c - 3.929).$$

The solution of this quadratic equation is $c = 197.274$. This in turn gives $a = 0.0313$. Therefore

$$P(t) = \frac{197.274}{1 + 49.21e^{-0.0313t}} .$$

(b)

Year	Census Population	Predicted Population	Error	% Error
1790	3.929	3.929	0.000	0.00
1800	5.308	5.334	-0.026	-0.49
1810	7.240	7.222	0.018	0.24
1820	9.638	9.746	-0.108	-1.12
1830	12.866	13.090	-0.224	-1.74
1840	17.069	17.475	-0.406	-2.38
1850	23.192	23.143	0.049	0.21
1860	31.433	30.341	1.092	3.47
1870	38.558	39.272	-0.714	-1.85
1880	50.156	50.044	0.112	0.22
1890	62.948	62.600	0.348	0.55
1900	75.996	76.666	-0.670	-0.88
1910	91.972	91.739	0.233	0.25
1920	105.711	107.143	-1.432	-1.35
1930	122.775	122.140	0.635	0.52
1940	131.669	136.068	-4.399	-3.34
1950	150.697	148.445	2.252	1.49

17

21. (a) Writing the equation in the form $(x - \sqrt{x^2 + y^2})dx + y\,dy$ we identify $M = x - \sqrt{x^2 + y^2}$ and $N = y$. Since M and N are both homogeneous of degree 1 we use the substitution $y = ux$. It follows that

$$\left(x - \sqrt{x^2 + u^2 x^2}\right) dx + ux(u\,dx + x\,du) = 0$$

$$x\left[\left(1 - \sqrt{1 + u^2}\right) + u^2\right] dx + x^2 u\,du = 0$$

$$-\frac{u\,du}{1 + u^2 - \sqrt{1 + u^2}} = \frac{dx}{x}$$

$$\frac{u\,du}{\sqrt{1 + u^2}\,(1 - \sqrt{1 + u^2})} = \frac{dx}{x}.$$

Letting $w = 1 - \sqrt{1 + u^2}$ we have $dw = -u\,du/\sqrt{1 + u^2}$ so that

$$-\ln\left(1 - \sqrt{1 + u^2}\right) = \ln x + c$$

$$\frac{1}{1 - \sqrt{1 + u^2}} = c_1 x$$

$$1 - \sqrt{1 + u^2} = -\frac{c_2}{x} \qquad (-c_2 = 1/c_1)$$

$$1 + \frac{c_2}{x} = \sqrt{1 + \frac{y^2}{x^2}}$$

$$1 + \frac{2c_2}{x} + \frac{c_2^2}{x^2} = 1 + \frac{y^2}{x^2}.$$

Solving for y^2 we have

$$y^2 = 2c_2 x + c_2^2 = 4\left(\frac{c_2}{2}\right)\left(x + \frac{c_2}{2}\right)$$

which is a family of parabolas symmetric with respect to the x-axis with vertex at $(-c_2/2, 0)$ and focus at the origin.

(b) Writing the differential equation as $yy' + x = \sqrt{x^2 + y^2}$ and then squaring and simplifying we obtain $y = 2xy' + y(y')^2$. Let $w = y^2$ and write the differential equation as $y^2 = 2xyy' + y^2(y')^2$. Now $dw/dx = 2yy'$ so $w = xw' + \frac{1}{4}(w')^2$. Using Problem 54 in Exercises 1.1 we obtain $w = cx + \frac{1}{4}c^2$ or $y^2 = cx + (c/2)^2$. Letting $c/2 = c_2$ we have $y^2 = 2c_2 x + c_2^2$, which is the solution obtained in part (a).

(c) Let $u = x^2 + y^2$ so that

$$\frac{du}{dx} = 2x + 2y\frac{dy}{dx}.$$

18

Then
$$y\frac{dy}{dx} = \frac{1}{2}\frac{du}{dx} - x$$

and the differential equation can be written in the form

$$\frac{1}{2}\frac{du}{dx} - x = -x + \sqrt{u} \quad \text{or} \quad \frac{1}{2}\frac{du}{dx} = \sqrt{u}.$$

Separating variables and integrating we have

$$\frac{du}{2\sqrt{u}} = dx$$

$$\sqrt{u} = x + c$$

$$u = x^2 + 2cx + c^2$$

$$x^2 + y^2 = x^2 + 2cx + c^2$$

$$y^2 = 2cx + c^2.$$

Exercises 3.3

3. The amounts of x and y are the same at about $t = 5$ days. The amounts of x and z are the same at about $t = 20$ days. The amounts of y and z are the same at about $t = 147$ days. The time when y and z are the same makes sense because most of A and half of B are gone, so half of C should have been formed.

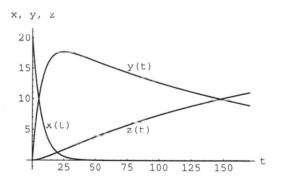

6. Let x_1, x_2, and x_3 be the amounts of salt in tanks A, B, and C, respectively, so that

$$x_1' = \frac{1}{100}x_2 \cdot 2 - \frac{1}{100}x_1 \cdot 6 = \frac{1}{50}x_2 - \frac{3}{50}x_1$$

$$x_2' = \frac{1}{100}x_1 \cdot 6 + \frac{1}{100}x_3 - \frac{1}{100}x_2 \cdot 2 - \frac{1}{100}x_2 \cdot 5 = \frac{3}{50}x_1 - \frac{7}{100}x_2 + \frac{1}{100}x_3$$

$$x_3' = \frac{1}{100}x_2 \cdot 5 - \frac{1}{100}x_3 - \frac{1}{100}x_3 \cdot 4 = \frac{1}{20}x_2 - \frac{1}{20}x_3.$$

9. From the graph we see that the populations are first equal at about $t = 5.6$. The approximate periods of x and y are both 45.

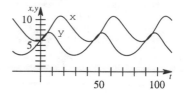

12. By Kirchoff's first law we have $i_1 = i_2 + i_3$. By Kirchoff's second law, on each loop we have $E(t) = Li_1' + R_1 i_2$ and $E(t) = Li_1' + R_2 i_3 + \frac{1}{C} q$ so that $q = CR_1 i_2 - CR_2 i_3$. Then $i_3 = q' = CR_1 i_2' - CR_2 i_3$ so that the system is

$$Li_2' + Li_3' + R_1 i_2 = E(t)$$

$$-R_1 i_2' + R_2 i_3' + \frac{1}{C} i_3 = 0.$$

15. We first note that $s(t) + i(t) + r(t) = n$. Now the rate of change of the number of susceptible persons, $s(t)$, is proportional to the number of contacts between the number of people infected and the number who are susceptible; that is, $ds/dt = -k_1 si$. We use $-k_1$ because $s(t)$ is decreasing. Next, the rate of change of the number of persons who have recovered is proportional to the number infected; that is, $dr/dt = k_2 i$ where k_2 is positive since r is increasing. Finally, to obtain di/dt we use

$$\frac{d}{dt}(s + i + r) = \frac{d}{dt} n = 0.$$

This gives

$$\frac{di}{dt} = -\frac{dr}{dt} - \frac{ds}{dt} = -k_2 i + k_1 si.$$

The system of equations is then

$$\frac{ds}{dt} = -k_1 si$$

$$\frac{di}{dt} = -k_2 i + k_1 si$$

$$\frac{dr}{dt} = k_2 i.$$

A reasonable set of initial conditions is $i(0) = i_0$, the number of infected people at time 0, $s(0) = n - i_0$, and $r(0) = 0$.

20

Chapter 3 Review Exercises

3. From $dE/dt = -E/RC$ and $E(t_1) = E_0$ we obtain $E = E_0 e^{(t_1-t)/RC}$.

6. We first solve $\left(1 - \dfrac{t}{10}\right)\dfrac{di}{dt} + 0.2i = 4$ Separating variables we obtain

$\dfrac{di}{40 - 2i} = \dfrac{dt}{10 - t}$. Then

$$-\frac{1}{2}\ln|40 - 2i| = -\ln|10 - t| + c \quad \text{or} \quad \sqrt{40 - 2i} = c_1(10 - t).$$

Since $i(0) = 0$ we must have $c_1 = 2/\sqrt{10}$. Solving for i we get $i(t) = 4t - \frac{1}{5}t^2$,

$0 \le t < 10$. For $t \ge 10$ the equation for the current becomes $0.2i = 4$ or $i = 20$. Thus

$$i(t) = \begin{cases} 4t - \frac{1}{5}t^2, & 0 \le t < 10 \\ 20, & t \ge 10 \end{cases}.$$

9. From $\dfrac{dx}{dt} = k_1 x(\alpha - x)$ we obtain $\left(\dfrac{1/\alpha}{x} + \dfrac{1/\alpha}{\alpha - x}\right) dx = k_1\,dt$ so that $x = \dfrac{\alpha c_1 e^{\alpha k_1 t}}{1 + c_1 e^{\alpha k_1 t}}$. From

$\dfrac{dy}{dt} = k_2 xy$ we obtain

$$\ln|y| = \frac{k_2}{k_1}\ln\left|1 + c_1 e^{\alpha k_1 t}\right| + c \quad \text{or} \quad y = c_2\left(1 + c_1 e^{\alpha k_1 t}\right)^{k_2/k_1}.$$

4 Differential Equations of Higher-Order

3. From $y = c_1 e^{4x} + c_2 e^{-x}$ we find $y' = 4c_1 e^{4x} - c_2 e^{-x}$. Then $y(0) = c_1 + c_2 = 1$, $y'(0) = 4c_1 - c_2 = 2$ so that $c_1 = 3/5$ and $c_2 = 2/5$. The solution is $y = \frac{3}{5} e^{4x} + \frac{2}{5} e^{-x}$.

6. From $y = c_1 + c_2 x^2$ we find $y' = 2c_2 x$. Then $y(0) = c_1 = 0$, $y'(0) = 2c_2 \cdot 0 = 0$ and $y'(0) = 1$ is not possible. Since $a_2(x) = x$ is 0 at $x = 0$, Theorem 4.1 is not violated.

9. From $y = c_1 e^x \cos x + c_2 e^x \sin x$ we find $y' = c_1 e^x (-\sin x + \cos x) + c_2 e^x (\cos x + \sin x)$.

  (a) We have $y(0) = c_1 = 1$, $y'(0) - c_1 + c_2 = 0$ so that $c_1 = 1$ and $c_2 - -1$. The solution is $y = e^x \cos x - e^x \sin x$.

  (b) We have $y(0) = c_1 = 1$, $y(\pi) = -c_1 e^\pi = -1$, which is not possible.

  (c) We have $y(0) = c_1 = 1$, $y(\pi/2) = c_2 e^{\pi/2} = 1$ so that $c_1 = 1$ and $c_2 = e^{-\pi/2}$. The solution is $y = e^x \cos x + e^{-\pi/2} e^x \sin x$.

  (d) We have $y(0) = c_1 = 0$, $y(\pi) = -c_1 e^\pi = 0$ so that $c_1 = 0$ and c_2 is arbitrary. Solutions are $y = c_2 e^x \sin x$, for any real numbers c_2.

12. Since $a_0(x) = \tan x$ and $x_0 = 0$ the problem has a unique solution for $-\pi/2 < x < \pi/2$.

15. Since $(-4)x + (3)x^2 + (1)(4x - 3x^2) = 0$ the functions are linearly dependent.

18. Since $(1)\cos 2x + (1)1 + (-2)\cos^2 x = 0$ the functions are linearly dependent.

21. The functions are linearly independent since $W\left(1 + x, x, x^2\right) = \begin{vmatrix} 1+x & x & x^2 \\ 1 & 1 & 2x \\ 0 & 0 & 2 \end{vmatrix} = 2 \neq 0.$

24. The functions satisfy the differential equation and are linearly independent since

$$W(\cosh 2x, \sinh 2x) = 2$$

for $-\infty < x < \infty$. The general solution is

$$y = c_1 \cosh 2x + c_2 \sinh 2x.$$

27. The functions satisfy the differential equation and are linearly independent since

$$W\left(x^3, x^4\right) = x^6 \neq 0$$

for $0 < x < \infty$. The general solution is

$$y = c_1 x^3 + c_2 x^4.$$

30. The functions satisfy the differential equation and are linearly independent since

$$W(1, x, \cos x, \sin x) = 1$$

for $-\infty < x < \infty$. The general solution is

$$y = c_1 + c_2 x + c_3 \cos x + c_4 \sin x.$$

33. The functions $y_1 = e^{2x}$ and $y_2 = e^{5x}$ form a fundamental set of solutions of the homogeneous equation, and $y_p = 6e^x$ is a particular solution of the nonhomogeneous equation.

36. The functions $y_1 = x^{-1/2}$ and $y_2 = x^{-1}$ form a fundamental set of solutions of the homogeneous equation, and $y_p = \frac{1}{15}x^2 - \frac{1}{6}x$ is a particular solution of the nonhomogeneous equation.

—————— **Exercises 4.2** ——————————

In Problems 3-9 we use reduction of order to find a second solution. In Problems 12-24 we use formula (5) from the text.

3. Define $y = u(x)e^{2x}$ so

$$y' = 2ue^{2x} + u'e^{2x}, \quad y'' = e^{2x}u'' + 4e^{2x}u' + 4e^{2x}u, \quad \text{and} \quad y'' - 4y' + 4y = 4e^{2x}u'' = 0.$$

Therefore $u'' = 0$ and $u = c_1 x + c_2$. Taking $c_1 = 1$ and $c_2 = 0$ we see that a second solution is $y_2 = xe^{2x}$.

6. Define $y = u(x)\sin 3x$ so

$$y' = 3u\cos 3x + u'\sin 3x, \quad y'' = u''\sin 3x + 6u'\cos 3x - 9u\sin 3x,$$

and

$$y'' + 9y = (\sin 3x)u'' + 6(\cos 3x)u' = 0 \quad \text{or} \quad u'' + 6(\cot 3x)u' = 0.$$

If $w = u'$ we obtain the first-order equation $w' + 6(\cot 3x)w = 0$ which has the integrating factor $e^{6\int \cot 3x\, dx} = \sin^2 3x$. Now

$$\frac{d}{dx}[(\sin^2 3x)w] = 0 \quad \text{gives} \quad (\sin^2 3x)w = c.$$

Therefore $w = u' = c\csc^2 3x$ and $u = c_1 \cot 3x$. A second solution is $y_2 = \cot 3x \sin 3x = \cos 3x$.

9. Define $y = u(x)e^{2x/3}$ so

$$y' = \frac{2}{3}e^{2x/3}u + e^{2x/3}u', \quad y'' = e^{2x/3}u'' + \frac{4}{3}e^{2x/3}u' + \frac{4}{9}e^{2x/3}u$$

and

$$9y'' - 12y' + 4y = 9e^{2x/3}u'' = 0.$$

23

Exercises 4.2

Therefore $u'' = 0$ and $u = c_1 x + c_2$. Taking $c_1 = 1$ and $c_2 = 0$ we see that a second solution is $y_2 = xe^{2x/3}$.

12. Identifying $P(x) = 2/x$ we have

$$y_2 = x^2 \int \frac{e^{-\int (2/x)\,dx}}{x^4}\,dx = x^2 \int x^{-6}\,dx = -\frac{1}{5}x^{-3}.$$

A second solution is $y_2 = x^{-3}$.

15. Identifying $P(x) = 2(1+x)/\left(1 - 2x - x^2\right)$ we have

$$y_2 = (x+1) \int \frac{e^{-\int 2(1+x)dx/\left(1-2x-x^2\right)}}{(x+1)^2}\,dx = (x+1) \int \frac{e^{\ln\left(1-2x-x^2\right)}}{(x+1)^2}\,dx$$

$$= (x+1) \int \frac{1 - 2x - x^2}{(x+1)^2}\,dx = (x+1) \int \left[\frac{2}{(x+1)^2} - 1\right]dx$$

$$= (x+1)\left[-\frac{2}{x+1} - x\right] = -2 - x^2 - x.$$

A second solution is $y_2 = x^2 + x + 2$.

18. Identifying $P(x) = -3/x$ we have

$$y_2 = x^2 \cos(\ln x) \int \frac{e^{-\int -3\,dx/x}}{x^4 \cos^2(\ln x)}\,dx = x^2 \cos(\ln x) \int \frac{x^3}{x^4 \cos^2(\ln x)}\,dx$$

$$= x^2 \cos(\ln x)\tan(\ln x) = x^2 \sin(\ln x).$$

A second solution is $y_2 = x^2 \sin(\ln x)$.

21. Identifying $P(x) = -1/x$ we have

$$y_2 = x \int \frac{e^{-\int -dx/x}}{x^2}\,dx = x \int \frac{dx}{x} = x \ln |x|.$$

A second solution is $y_2 = x \ln |x|$.

24. Identifying $P(x) = 1/x$ we have

$$y_2 = \cos(\ln x) \int \frac{e^{-\int dx/x}}{\cos^2(\ln x)}\,dx = \cos(\ln x) \int \frac{1/x}{\cos^2(\ln x)}\,dx = \cos(\ln x)\tan(\ln x) = \sin(\ln x).$$

A second solution is $y_2 = \sin(\ln x)$.

27. Define $y = u(x)e^x$ so

$$y' = ue^x + u'e^x, \quad y'' = u''e^x + 2u'e^x + ue^x$$

and

$$y'' - 3y' + 2y = e^x u'' - e^x u' = 0 \quad \text{or} \quad u'' - u' = 0.$$

24

If $w = u'$ we obtain the first order equation $w' - w = 0$ which has the integrating factor $e^{-\int dx} = e^{-x}$. Now

$$\frac{d}{dx}[e^{-x}w] = 0 \quad \text{gives} \quad e^{-x}w = c.$$

Therefore $w = u' = ce^x$ and $u = ce^x$. A second solution is $y_2 = e^x e^x = e^{2x}$. To find a particular solution we try $y_p = Ae^{3x}$. Then $y' = 3Ae^{3x}$, $y'' = 9Ae^{3x}$, and $9Ae^{3x} - 3\left(3Ae^{3x}\right) + 2Ae^{3x} = 5e^{3x}$. Thus $A = 5/2$ and $y_p = \frac{5}{2}e^{3x}$. The general solution is

$$y = c_1 e^x + c_2 e^{2x} + \frac{5}{2}e^{3x}.$$

Exercises 4.3

3. From $m^2 - 36 = 0$ we obtain $m = 6$ and $m = -6$ so that $y = c_1 e^{6x} + c_2 e^{-6x}$.

6. From $3m^2 + 1 = 0$ we obtain $m = i/\sqrt{3}$ and $m = -i/\sqrt{3}$ so that $y = c_1 \cos x/\sqrt{3} + c_2 \sin x/\sqrt{3}$.

9. From $m^2 + 8m + 16 = 0$ we obtain $m = -4$ and $m = -4$ so that $y = c_1 e^{-4x} + c_2 x e^{-4x}$.

12. From $m^2 + 4m - 1 = 0$ we obtain $m = -2 \pm \sqrt{5}$ so that $y = c_1 e^{(-2+\sqrt{5})x} + c_2 e^{(-2-\sqrt{5})x}$.

15. From $m^2 - 4m + 5 = 0$ we obtain $m = 2 \pm i$ so that $y = e^{2x}(c_1 \cos x + c_2 \sin x)$.

18. From $2m^2 + 2m + 1 = 0$ we obtain $m = -1/2 \pm i/2$ so that

$$y = e^{-x/2}(c_1 \cos x/2 + c_2 \sin x/2).$$

21. From $m^3 - 1 = 0$ we obtain $m = 1$ and $m = -1/2 \pm \sqrt{3}\,i/2$ so that

$$y = c_1 e^x + e^{-x/2}\left(c_2 \cos \sqrt{3}\,x/2 + c_3 \sin \sqrt{3}\,x/2\right).$$

24. From $m^3 + 3m^2 - 4m - 12 = 0$ we obtain $m = -2$, $m = 2$, and $m = -3$ so that

$$y = c_1 e^{-2x} + c_2 e^{2x} + c_3 e^{-3x}.$$

27. From $m^3 + 3m^2 + 3m + 1 = 0$ we obtain $m = -1$, $m = -1$, and $m = -1$ so that

$$y = c_1 e^{-x} + c_2 x e^{-x} + c_3 x^2 e^{-x}.$$

30. From $m^4 - 2m^2 + 1 = 0$ we obtain $m = 1$, $m = 1$, $m = -1$, and $m = -1$ so that

$$y = c_1 e^x + c_2 x e^x + c_3 e^{-x} + c_4 x e^{-x}.$$

33. From $m^5 - 16m = 0$ we obtain $m = 0$, $m = 2$, $m = -2$, and $m = \pm 2i$ so that

$$y = c_1 + c_2 e^{2x} + c_3 e^{-2x} + c_4 \cos 2x + c_5 \sin 2x.$$

36. From $2m^5 - 7m^4 + 12m^3 + 8m^2 = 0$ we obtain $m = 0$, $m = 0$, $m = -1/2$, and $m = 2 \pm 2i$ so that

$$y = c_1 + c_2 x + c_3 e^{-x/2} + e^{2x}(c_4 \cos 2x + c_5 \sin 2x).$$

39. From $m^2 + 6m + 5 = 0$ we obtain $m = -1$ and $m = -5$ so that $y = c_1 e^{-x} + c_2 e^{-5x}$. If $y(0) = 0$ and $y'(0) = 3$ then $c_1 + c_2 = 0$, $-c_1 - 5c_2 = 3$, so $c_1 = 3/4$, $c_2 = -3/4$, and $y = \frac{3}{4}e^{-x} - \frac{3}{4}e^{-5x}$.

42. From $m^2 - 2m + 1 = 0$ we obtain $m = 1$ and $m = 1$ so that $y = c_1 e^x + c_2 x e^x$. If $y(0) = 5$ and $y'(0) = 10$ then $c_1 = 5$, $c_1 + c_2 = 10$ so $c_1 = 5$, $c_2 = 5$, and $y = 5e^x + 5xe^x$.

45. From $m^2 - 3m + 2 = 0$ we obtain $m = 1$ and $m = 2$ so that $y = c_1 e^x + c_2 e^{2x}$. If $y(1) = 0$ and $y'(1) = 1$ then $c_1 e + c_2 e^2 = 0$, $c_1 e + 2c_2 e^2 = 0$ so $c_1 = -e^{-1}$, $c_2 = e^{-2}$, and $y = -e^{x-1} + e^{2x-2}$.

48. From $m^3 + 2m^2 - 5m - 6 = 0$ we obtain $m = -1$, $m = 2$, and $m = -3$ so that

$$y = c_1 e^{-x} + c_2 e^{2x} + c_3 e^{-3x}.$$

If $y(0) = 0$, $y'(0) = 0$, and $y''(0) = 1$ then

$$c_1 + c_2 + c_3 = 0, \quad -c_1 + 2c_2 - 3c_3 = 0, \quad c_1 + 4c_2 + 9c_3 = 1,$$

so $c_1 = -1/6$, $c_2 = 1/15$, $c_3 = 1/10$, and

$$y = -\frac{1}{6}e^{-x} + \frac{1}{15}e^{2x} + \frac{1}{10}e^{-3x}.$$

51. From $m^4 - 3m^3 + 3m^2 - m = 0$ we obtain $m = 0$, $m = 1$, $m = 1$, and $m = 1$ so that $y = c_1 + c_2 e^x + c_3 x e^x + c_4 x^2 e^x$. If $y(0) = 0$, $y'(0) = 0$, $y''(0) = 1$, and $y'''(0) = 1$ then

$$c_1 + c_2 = 0, \quad c_2 + c_3 = 0, \quad c_2 + 2c_3 + 2c_4 = 1, \quad c_2 + 3c_3 + 6c_4 = 1,$$

so $c_1 = 2$, $c_2 = -2$, $c_3 = 2$, $c_4 = -1/2$, and

$$y = 2 - 2e^x + 2xe^x - \frac{1}{2}x^2 e^x.$$

54. From $m^2 + 4 = 0$ we obtain $m = \pm 2i$ so that $y = c_1 \cos 2x + c_2 \sin 2x$. If $y(0) = 0$ and $y(\pi) = 0$ then $c_1 = 0$ and $y = c_2 \sin 2x$.

57. Using a CAS to solve the auxiliary equation $m^3 - 6m^2 + 2m + 1$ we find $m_1 = -0.270534$, $m_2 = 0.658675$, and $m_3 = 5.61186$. The general solution is

$$y = c_1 e^{-0.270534x} + c_2 e^{0.658675x} + c_3 e^{5.61186x}.$$

60. Using a CAS to solve the auxiliary equation $m^4 + 2m^2 - m + 2 = 0$ we find $m_1 = 1/2 + \sqrt{3}\,i/2$, $m_2 = 1/2 - \sqrt{3}\,i/2$, $m_3 = -1/2 + \sqrt{7}\,i/2$, and $m_4 = -1/2 - \sqrt{7}\,i/2$. The general solution is

$$y = e^{x/2}\left(c_1 \cos\frac{\sqrt{3}}{2}x + c_2 \sin\frac{\sqrt{3}}{2}x\right) + e^{-x/2}\left(c_3 \cos\frac{\sqrt{7}}{2}x + c_4 \sin\frac{\sqrt{7}}{2}x\right).$$

3. From $m^2 - 10m + 25 = 0$ we find $m_1 = m_2 = 5$. Then $y_c = c_1 e^{5x} + c_2 x e^{5x}$ and we assume $y_p = Ax + B$. Substituting into the differential equation we obtain $25A = 30$ and $-10A + 25B = 3$. Then $A = \frac{6}{5}$, $B = \frac{6}{5}$, $y_p = \frac{6}{5}x + \frac{6}{5}$, and

$$y = c_1 e^{5x} + c_2 x e^{5x} + \frac{6}{5}x + \frac{6}{5}.$$

6. From $m^2 - 8m + 20 = 0$ we find $m_1 = 2 + 4i$ and $m_2 = 2 - 4i$. Then $y_c = e^{2x}(c_1 \cos 4x + c_2 \sin 4x)$ and we assume $y_p = Ax^2 + Bx + C + (Dx + E)e^x$. Substituting into the differential equation we obtain

$$2A - 8B + 20C = 0$$

$$-6D + 13E = 0$$

$$-16A + 20B = 0$$

$$13D = -26$$

$$20A = 100.$$

Then $A = 5$, $B = 4$, $C = \frac{11}{10}$, $D = -2$, $E = -\frac{12}{13}$, $y_p = 5x^2 + 4x + \frac{11}{10} + \left(-2x - \frac{12}{13}\right)e^x$ and

$$y = e^{2x}(c_1 \cos 4x + c_2 \sin 4x) + 5x^2 + 4x + \frac{11}{10} + \left(-2x - \frac{12}{13}\right)e^x.$$

9. From $m^2 - m = 0$ we find $m_1 = 1$ and $m_2 = 0$. Then $y_c = c_1 e^x + c_2$ and we assume $y_p = Ax$. Substituting into the differential equation we obtain $-A = -3$. Then $A = 3$, $y_p = 3x$ and $y = c_1 e^x + c_2 + 3x$.

12. From $m^2 - 16 = 0$ we find $m_1 = 4$ and $m_2 = -4$. Then $y_c = c_1 e^{4x} + c_2 e^{-4x}$ and we assume $y_p = Axe^{4x}$. Substituting into the differential equation we obtain $8A = 2$. Then $A = \frac{1}{4}$, $y_p = \frac{1}{4}xe^{4x}$ and

$$y = c_1 e^{4x} + c_2 e^{-4x} + \frac{1}{4}xe^{4x}.$$

15. From $m^2 + 1 = 0$ we find $m_1 = i$ and $m_2 = -i$. Then $y_c = c_1 \cos x + c_2 \sin x$ and we assume $y_p = (Ax^2 + Bx)\cos x + (Cx^2 + Dx)\sin x$. Substituting into the differential equation we obtain $4C = 0$, $2A + 2D = 0$, $-4A = 2$, and $-2B + 2C = 0$. Then $A = -\frac{1}{2}$, $B = 0$, $C = 0$, $D = \frac{1}{2}$, $y_p = -\frac{1}{2}x^2 \cos x + \frac{1}{2}x \sin x$, and

$$y = c_1 \cos x + c_2 \sin x - \frac{1}{2}x^2 \cos x + \frac{1}{2}x \sin x.$$

27

18. From $m^2 - 2m + 2 = 0$ we find $m_1 = 1 + i$ and $m_2 = 1 - i$. Then $y_c = e^x(c_1 \cos x + c_2 \sin x)$ and we assume $y_p = Ae^{2x} \cos x + Be^{2x} \sin x$. Substituting into the differential equation we obtain $A + 2B = 1$ and $-2A + B = -3$. Then $A = \frac{7}{5}$, $B = -\frac{1}{5}$, $y_p = \frac{7}{5}e^{2x} \cos x - \frac{1}{5}e^{2x} \sin x$ and

$$y = e^x(c_1 \cos x + c_2 \sin x) + \frac{7}{5}e^{2x} \cos x - \frac{1}{5}e^{2x} \sin x.$$

21. From $m^3 - 6m^2 = 0$ we find $m_1 = m_2 = 0$ and $m_3 = 6$. Then $y_c = c_1 + c_2 x + c_3 e^{6x}$ and we assume $y_p = Ax^2 + B \cos x + C \sin x$. Substituting into the differential equation we obtain $-12A = 3$, $6B - C = -1$, and $B + 6C = 0$. Then $A = -\frac{1}{4}$, $B = -\frac{6}{37}$, $C = \frac{1}{37}$, $y_p = -\frac{1}{4}x^2 - \frac{6}{37} \cos x + \frac{1}{37} \sin x$, and

$$y = c_1 + c_2 x + c_3 e^{6x} - \frac{1}{4}x^2 - \frac{6}{37} \cos x + \frac{1}{37} \sin x.$$

24. From $m^3 - m^2 - 4m + 4 = 0$ we find $m_1 = 1$, $m_2 = 2$, and $m_3 = -2$. Then $y_c = c_1 e^x + c_2 e^{2x} + c_3 e^{-2x}$ and we assume $y_p = A + Bxe^x + Cxe^{2x}$. Substituting into the differential equation we obtain $4A = 5$, $-3B = -1$, and $4C = 1$. Then $A = \frac{5}{4}$, $B = \frac{1}{3}$, $C = \frac{1}{4}$, $y_p = \frac{5}{4} + \frac{1}{3}xe^x + \frac{1}{4}xe^{2x}$, and

$$y = c_1 e^x + c_2 e^{2x} + c_3 e^{-2x} + \frac{5}{4} + \frac{1}{3}xe^x + \frac{1}{4}xe^{2x}.$$

27. We have $y_c = c_1 \cos 2x + c_2 \sin 2x$ and we assume $y_p = A$. Substituting into the differential equation we find $A = -\frac{1}{2}$. Thus $y = c_1 \cos 2x + c_2 \sin 2x - \frac{1}{2}$. From the initial conditions we obtain $c_1 = 0$ and $c_2 = \sqrt{2}$, so $y = \sqrt{2} \sin 2x - \frac{1}{2}$.

30. We have $y_c = c_1 e^{-2x} + c_2 xe^{-2x}$ and we assume $y_p = (Ax^3 + Bx^2)e^{-2x}$. Substituting into the differential equation we find $A = \frac{1}{6}$ and $B = \frac{3}{2}$. Thus $y = c_1 e^{-2x} + c_2 xe^{-2x} + \left(\frac{1}{6}x^3 + \frac{3}{2}x^2\right)e^{-2x}$. From the initial conditions we obtain $c_1 = 2$ and $c_2 = 9$, so

$$y = 2e^{-2x} + 9xe^{-2x} + \left(\frac{1}{6}x^3 + \frac{3}{2}x^2\right)e^{-2x}.$$

33. We have $x_c = c_1 \cos \omega t + c_2 \sin \omega t$ and we assume $x_p = At \cos \omega t + Bt \sin \omega t$. Substituting into the differential equation we find $A = -F_0/2\omega$ and $B = 0$. Thus $x = c_1 \cos \omega t + c_2 \sin \omega t - (F_0/2\omega)t \cos \omega t$. From the initial conditions we obtain $c_1 = 0$ and $c_2 = F_0/2\omega^2$, so

$$x = (F_0/2\omega^2) \sin \omega t - (F_0/2\omega)t \cos \omega t.$$

36. We have $y_c = c_1 e^{-2x} + e^x(c_2 \cos \sqrt{3}\,x + c_3 \sin \sqrt{3}\,x)$ and we assume $y_p = Ax + B + Cxe^{-2x}$. Substituting into the differential equation we find $A = \frac{1}{4}$, $B = -\frac{5}{8}$, and $C = \frac{2}{3}$. Thus

$$y = c_1 e^{-2x} + e^x(c_2 \cos \sqrt{3}\,x + c_3 \sin \sqrt{3}\,x) + \frac{1}{4}x - \frac{5}{8} + \frac{2}{3}xe^{-2x}.$$

From the initial conditions we obtain $c_1 = -\frac{23}{12}$, $c_2 = -\frac{59}{24}$, and $c_3 = \frac{17}{72}\sqrt{3}$, so

$$y = -\frac{23}{12}e^{-2x} + e^x\left(-\frac{59}{24} \cos \sqrt{3}\,x + \frac{17}{72}\sqrt{3} \sin \sqrt{3}\,x\right) + \frac{1}{4}x - \frac{5}{8} + \frac{2}{3}xe^{-2x}.$$

39. We have $y_c = c_1 \cos 2x + c_2 \sin 2x$ and we assume $y_p = A \cos x + B \sin x$ on $[0, \pi/2]$. Substituting into the differential equation we find $A = 0$ and $B = \frac{1}{3}$. Thus $y = c_1 \cos 2x + c_2 \sin 2x + \frac{1}{3} \sin x$ on $[0, \pi/2]$. On $(\pi/2, \infty)$ we have $y = c_3 \cos 2x + c_4 \sin 2x$. From $y(0) = 1$ and $y'(0) = 2$ we obtain

$$c_1 = 1$$

$$\frac{1}{3} + 2c_2 = 2.$$

Solving this system we find $c_1 = 1$ and $c_2 = \frac{5}{6}$. Thus $y = \cos 2x + \frac{5}{6} \sin 2x + \frac{1}{3} \sin x$ on $[0, \pi/2]$. Now continuity of y at $x = \pi/2$ implies

$$\cos \pi + \frac{5}{6} \sin \pi + \frac{1}{3} \sin \frac{\pi}{2} = c_3 \cos \pi + c_4 \sin \pi$$

or $-1 + \frac{1}{3} = -c_3$. Hence $c_3 = \frac{2}{3}$. Continuity of y' at $x = \pi/2$ implies

$$-2 \sin \pi + \frac{5}{3} \cos \pi + \frac{1}{3} \cos \frac{\pi}{2} = -2c_3 \sin \pi + 2c_4 \cos \pi$$

or $-\frac{5}{3} = -2c_4$. Then $c_4 = \frac{5}{6}$ and the solution of the boundary-value problem is

$$y(x) = \begin{cases} \cos 2x + \frac{5}{6} \sin 2x + \frac{1}{3} \sin x, & 0 \le x \le \pi/2 \\ \frac{2}{3} \cos 2x + \frac{5}{6} \sin 2x, & x > \pi/2 \end{cases}$$

Exercises 4.5

3. $(D^2 - 4D - 12)y = (D - 6)(D + 2)y = x - 6$

6. $(D^3 + 4D)y = D(D^2 + 4)y = e^x \cos 2x$

9. $(D^4 + 8D)y = D(D + 2)(D^2 - 2D + 4)y = 4$

12. $(2D - 1)y = (2D - 1)4e^{x/2} = 8De^{x/2} - 4e^{x/2} = 4e^{x/2} - 4e^{x/2} = 0$

15. D^4 because of x^3

18. $D^2(D - 6)^2$ because of x and xe^{6x}

21. $D^3(D^2 + 16)$ because of x^2 and $\sin 4x$

24. $D(D - 1)(D - 2)$ because of 1, e^x, and e^{2x}

27. $1, x, x^2, x^3, x^4$

30. $D^2 - 9D - 36 = (D - 12)(D + 3);\quad e^{12x}, e^{-3x}$

33. $D^3 - 10D^2 + 25D = D(D - 5)^2;\quad 1, e^{5x}, xe^{5x}$

36. Applying D to the differential equation we obtain

$$D(2D^2 - 7D + 5)y = 0.$$

Then

$$y = \underbrace{c_1 e^{5x/2} + c_2 e^x}_{y_c} + c_3$$

and $y_p = A$. Substituting y_p into the differential equation yields $5A = -29$ or $A = -29/5$. The general solution is

$$y = c_1 e^{5x/2} + c_2 e^x - \frac{29}{5}.$$

39. Applying D^2 to the differential equation we obtain

$$D^2(D^2 + 4D + 4)y = D^2(D + 2)^2 y = 0.$$

Then

$$y = \underbrace{c_1 e^{-2x} + c_2 x e^{-2x}}_{y_c} + c_3 + c_4 x$$

and $y_p = Ax + B$. Substituting y_p into the differential equation yields $4Ax + (4A + 4B) = 2x + 6$. Equating coefficients gives

$$4A = 2$$

$$4A + 4B = 6.$$

Then $A = 1/2$, $B = 1$, and the general solution is

$$y = c_1 e^{-2x} + c_2 x e^{-2x} + \frac{1}{2}x + 1.$$

42. Applying D^4 to the differential equation we obtain

$$D^4(D^2 - 2D + 1)y = D^4(D - 1)^2 y = 0.$$

Then

$$y = \underbrace{c_1 e^x + c_2 x e^x}_{y_c} + c_3 x^3 + c_4 x^2 + c_5 x + c_6$$

and $y_p = Ax^3 + Bx^2 + Cx + D$. Substituting y_p into the differential equation yields
$Ax^3 + (B - 6A)x^2 + (6A - 4B + C)x + (2B - 2C + D) = x^3 + 4x$. Equating coefficients gives

$$A = 1$$

$$B - 6A = 0$$

$$6A - 4B + C = 4$$

$$2B - 2C + D = 0.$$

Then $A = 1$, $B = 6$, $C = 22$, $D = 32$, and the general solution is

$$y = c_1 e^x + c_2 x e^x + x^3 + 6x^2 + 22x + 32.$$

30

45. Applying $D(D-1)$ to the differential equation we obtain

$$D(D-1)(D^2 - 2D - 3)y = D(D-1)(D+1)(D-3)y = 0.$$

Then

$$y = \underbrace{c_1 e^{3x} + c_2 e^{-x}}_{y_c} + c_3 e^{x} + c_4$$

and $y_p = Ae^x + B$. Substituting y_p into the differential equation yields $-4Ae^x - 3B = 4e^x - 9$. Equating coefficients gives $A = -1$ and $B = 3$. The general solution is

$$y = c_1 e^{3x} + c_2 e^{-x} - e^x + 3.$$

48. Applying $D(D^2 + 1)$ to the differential equation we obtain

$$D(D^2 + 1)(D^2 + 4)y = 0.$$

Then

$$y = \underbrace{c_1 \cos 2x + c_2 \sin 2x}_{y_c} + c_3 \cos x + c_4 \sin x + c_5$$

and $y_p = A \cos x + B \sin x + C$. Substituting y_p into the differential equation yields $3A \cos x + 3B \sin x + 4C = 4 \cos x + 3 \sin x - 8$. Equating coefficients gives $A = 4/3$, $B = 1$, and $C = -2$. The general solution is

$$y = c_1 \cos 2x + c_2 \sin 2x + \frac{4}{3} \cos x + \sin x - 2.$$

51. Applying $D(D-1)^3$ to the differential equation we obtain

$$D(D-1)^3(D^2 - 1)y = D(D-1)^4(D+1)y = 0.$$

Then

$$y = \underbrace{c_1 e^x + c_2 e^{-x}}_{y_c} + c_3 x^3 e^x + c_4 x^2 e^x + c_5 x e^x + c_6$$

and $y_p = Ax^3 e^x + Bx^2 e^x + Cx e^x + D$. Substituting y_p into the differential equation yields $6Ax^2 e^x + (6A + 4B)xe^x + (2B + 2C)e^x - D = x^2 e^x + 5$. Equating coefficients gives

$$6A = 1$$

$$6A + 4B = 0$$

$$2B + 2C = 0$$

$$-D = 5.$$

Then $A = 1/6$, $B = -1/4$, $C = 1/4$, $D = -5$, and the general solution is

$$y = c_1 e^x + c_2 e^{-x} + \frac{1}{6} x^3 e^x - \frac{1}{4} x^2 e^x + \frac{1}{4} x e^x - 5.$$

31

Exercises 4.5

54. Applying $D^2 - 2D + 10$ to the differential equation we obtain

$$(D^2 - 2D + 10)\left(D^2 + D + \frac{1}{4}\right)y = (D^2 - 2D + 10)\left(D + \frac{1}{2}\right)^2 y = 0.$$

Then

$$y = \underbrace{c_1 e^{-x/2} + c_2 x e^{-x/2}}_{y_c} + c_3 e^x \cos 3x + c_4 e^x \sin 3x$$

and $y_p = Ae^x \cos 3x + Be^x \sin 3x$. Substituting y_p into the differential equation yields $(9B - 27A/4)e^x \cos 3x - (9A + 27B/4)e^x \sin 3x = -e^x \cos 3x + e^x \sin 3x$. Equating coefficients gives

$$-\frac{27}{4}A + 9B = -1$$

$$-9A - \frac{27}{4}B = 1.$$

Then $A = -4/225$, $B = -28/225$, and the general solution is

$$y = c_1 e^{-x/2} + c_2 x e^{-x/2} - \frac{4}{225}e^x \cos 3x - \frac{28}{225}e^x \sin 3x.$$

57. Applying $(D^2 + 1)^2$ to the differential equation we obtain

$$(D^2 + 1)^2(D^2 + D + 1) = 0.$$

Then

$$y = \underbrace{e^{-x/2}\left[c_1 \cos \frac{\sqrt{3}}{2}x + c_2 \sin \frac{\sqrt{3}}{2}x\right]}_{y_c} + c_3 \cos x + c_4 \sin x + c_5 x \cos x + c_6 x \sin x$$

and $y_p = A \cos x + B \sin x + Cx \cos x + Dx \sin x$. Substituting y_p into the differential equation yields

$$(B + C + 2D)\cos x + Dx \cos x + (-A - 2C + D)\sin x - Cx \sin x = x \sin x.$$

Equating coefficients gives

$$B + C + 2D = 0$$

$$D = 0$$

$$-A - 2C + D = 0$$

$$-C = 1.$$

Then $A = 2$, $B = 1$, $C = -1$, and $D = 0$, and the general solution is

$$y = e^{-x/2}\left[c_1 \cos \frac{\sqrt{3}}{2}x + c_2 \sin \frac{\sqrt{3}}{2}x\right] + 2\cos x + \sin x - x \cos x.$$

32

60. Applying $D(D-1)^2(D+1)$ to the differential equation we obtain

$$D(D-1)^2(D+1)(D^3-D^2+D-1) = D(D-1)^3(D+1)(D^2+1) = 0.$$

Then

$$y = \underbrace{c_1 e^x + c_2 \cos x + c_3 \sin x}_{y_c} + c_4 + c_5 e^{-x} + c_6 x e^x + c_7 x^2 e^x$$

and $y_p = A + Be^{-x} + Cxe^x + Dx^2 e^x$. Substituting y_p into the differential equation yields

$$4Dxe^x + (2C+4D)e^x - 4Be^{-x} - A = xe^x - e^{-x} + 7.$$

Equating coefficients gives

$$4D = 1$$

$$2C + 4D = 0$$

$$-4B = -1$$

$$-A = 7.$$

Then $A = -7$, $B = 1/4$, $C = -1/2$, and $D = 1/4$, and the general solution is

$$y = c_1 e^x + c_2 \cos x + c_3 \sin x - 7 + \frac{1}{4}e^{-x} - \frac{1}{2}xe^x + \frac{1}{4}x^2 e^x.$$

63. Applying $D(D-1)$ to the differential equation we obtain

$$D(D-1)(D^4 - 2D^3 + D^2) = D^3(D-1)^3 = 0.$$

Then

$$y = \underbrace{c_1 + c_2 x + c_3 e^x + c_4 x e^x}_{y_c} + c_5 x^2 + c_6 x^2 e^x$$

and $y_p = Ax^2 + Bx^2 e^x$. Substituting y_p into the differential equation yields $2A + 2Be^x = 1 + e^x$. Equating coefficients gives $A = 1/2$ and $B = 1/2$. The general solution is

$$y = c_1 + c_2 x + c_3 e^x + c_4 x e^x + \frac{1}{2}x^2 + \frac{1}{2}x^2 e^x.$$

66. The complementary function is $y_c = c_1 + c_2 e^{-x}$. Using D^2 to annihilate x we find $y_p = Ax + Bx^2$. Substituting y_p into the differential equation we obtain $(A + 2B) + 2Bx = x$. Thus $A = -1$ and $B = 1/2$, and

$$y = c_1 + c_2 e^{-x} - x + \frac{1}{2}x^2$$

$$y' = -c_2 e^{-x} - 1 + x.$$

The initial conditions imply

$$c_1 + c_2 = 1$$

$$-c_2 = 1.$$

Thus $c_1 = 2$ and $c_2 = -1$, and

$$y = 2 - e^{-x} - x + \frac{1}{2}x^2.$$

69. The complementary function is $y_c = c_1 \cos x + c_2 \sin x$. Using $(D^2 + 1)(D^2 + 4)$ to annihilate $8 \cos 2x - 4 \sin x$ we find $y_p = Ax \cos x + Bx \sin x + C \cos 2x + D \sin 2x$. Substituting y_p into the differential equation we obtain $2B \cos x - 3C \cos 2x - 2A \sin x - 3D \sin 2x = 8 \cos 2x - 4 \sin x$. Thus $A = 2$, $B = 0$, $C = -8/3$, and $D = 0$, and

$$y = c_1 \cos x + c_2 \sin x + 2x \cos x - \frac{8}{3} \cos 2x$$

$$y' = -c_1 \sin x + c_2 \cos x + 2 \cos x - 2x \sin x + \frac{16}{3} \sin 2x.$$

The initial conditions imply

$$c_2 + \frac{8}{3} = -1$$

$$-c_1 - \pi = 0.$$

Thus $c_1 = -\pi$ and $c_2 = -11/3$, and

$$y = -\pi \cos x - \frac{11}{3} \sin x + 2x \cos x - \frac{8}{3} \cos 2x.$$

72. The complementary function is $y_c = c_1 + c_2 x + c_3 x^2 + c_4 e^x$. Using $D^2(D - 1)$ to annihilate $x + e^x$ we find $y_p = Ax^3 + Bx^4 + Cxe^x$. Substituting y_p into the differential equation we obtain $(-6A + 24B) - 24Bx + Ce^x = x + e^x$. Thus $A = -1/6$, $B = -1/24$, and $C = 1$, and

$$y = c_1 + c_2 x + c_3 x^2 + c_4 e^x - \frac{1}{6}x^3 - \frac{1}{24}x^4 + xe^x$$

$$y' = c_2 + 2c_3 x + c_4 e^x - \frac{1}{2}x^2 - \frac{1}{6}x^3 + e^x + xe^x$$

$$y'' = 2c_3 + c_4 e^x - x - \frac{1}{2}x^2 + 2e^x + xe^x.$$

$$y''' = c_4 e^x - 1 - x + 3e^x + xe^x$$

34

The initial conditions imply

$$c_1 + c_4 = 0$$

$$c_2 + c_4 + 1 = 0$$

$$2c_3 + c_4 + 2 = 0$$

$$2 + c_4 = 0.$$

Thus $c_1 = 2$, $c_2 = 1$, $c_3 = 0$, and $c_4 = -2$, and

$$y = 2 + x - 2e^x - \frac{1}{6}x^3 - \frac{1}{24}x^4 + xe^x.$$

———————— **Exercises 4.6** ————————————————————

The particular solution, $y_p = u_1 y_1 + u_2 y_2$, in the following problems can take on a variety of forms, especially where trigonometric functions are involved. The validity of a particular form can best be checked by substituting it back into the differential equation.

3. The auxiliary equation is $m^2 + 1 = 0$, so $y_c = c_1 \cos x + c_2 \sin x$ and

$$W = \begin{vmatrix} \cos x & \sin x \\ -\sin x & \cos x \end{vmatrix} = 1.$$

Identifying $f(x) = \sin x$ we obtain

$$u_1' = -\sin^2 x$$

$$u_2' - \cos x \sin x.$$

Then

$$u_1 = \frac{1}{4}\sin 2x - \frac{1}{2}x = \frac{1}{2}\sin x \cos x - \frac{1}{2}x$$

$$u_2 = -\frac{1}{2}\cos^2 x.$$

and

$$y = c_1 \cos x + c_2 \sin x + \frac{1}{2}\sin x \cos^2 x - \frac{1}{2}x \cos x - \frac{1}{2}\cos^2 x \sin x$$

$$= c_1 \cos x + c_2 \sin x - \frac{1}{2}x \cos x$$

for $-\infty < x < \infty$.

6. The auxiliary equation is $m^2 + 1 = 0$, so $y_c = c_1 \cos x + c_2 \sin x$ and

$$W = \begin{vmatrix} \cos x & \sin x \\ -\sin x & \cos x \end{vmatrix} = 1.$$

Identifying $f(x) = \sec^2 x$ we obtain

$$u_1' = -\frac{\sin x}{\cos^2 x}$$

$$u_2' = \sec x.$$

Then

$$u_1 = -\frac{1}{\cos x} = -\sec x$$

$$u_2 = \ln|\sec x + \tan x|$$

and

$$y = c_1 \cos x + c_2 \sin x - \cos x \sec x + \sin x \ln|\sec x + \tan x|$$

$$= c_1 \cos x + c_2 \sin x - 1 + \sin x \ln|\sec x + \tan x|$$

for $-\pi/2 < x < \pi/2$.

9. The auxiliary equation is $m^2 - 4 = 0$, so $y_c = c_1 e^{2x} + c_2 e^{-2x}$ and

$$W = \begin{vmatrix} e^{2x} & e^{-2x} \\ 2e^{2x} & -2e^{-2x} \end{vmatrix} = -4.$$

Identifying $f(x) = e^{2x}/x$ we obtain $u_1' = 1/4x$ and $u_2' = -e^{4x}/4x$. Then

$$u_1 = \frac{1}{4}\ln|x|, \qquad u_2 = -\frac{1}{4}\int_{x_0}^{x} \frac{e^{4t}}{t}\, dt$$

and

$$y = c_1 e^{2x} + c_2 e^{-2x} + \frac{1}{4}\left(e^{2x}\ln|x| - e^{-2x}\int_{x_0}^{x} \frac{e^{4t}}{t}\, dt \right), \qquad x_0 > 0$$

for $x > 0$.

12. The auxiliary equation is $m^2 - 3m + 2 = (m-1)(m-2) = 0$, so $y_c = c_1 e^x + c_2 e^{2x}$ and

$$W = \begin{vmatrix} e^x & e^{2x} \\ e^x & 2e^{2x} \end{vmatrix} = e^{3x}.$$

Identifying $f(x) = e^{3x}/(1+e^x)$ we obtain

$$u_1' = -\frac{e^{2x}}{1+e^x} = \frac{e^x}{1+e^x} - e^x$$

$$u_2' = \frac{e^x}{1+e^x}.$$

36

Then $u_1 = \ln(1 + e^x) - e^x$, $u_2 = \ln(1 + e^x)$, and

$$y = c_1 e^x + c_2 e^{2x} + e^x \ln(1 + e^x) - e^{2x} + e^{2x} \ln(1 + e^x)$$

$$= c_1 e^x + c_3 e^{2x} + (1 + e^x) e^x \ln(1 + e^x)$$

for $-\infty < x < \infty$.

15. The auxiliary equation is $m^2 - 2m + 1 = (m-1)^2 = 0$, so $y_c = c_1 e^x + c_2 x e^x$ and

$$W = \begin{vmatrix} e^x & x e^x \\ e^x & x e^x + e^x \end{vmatrix} = e^{2x}.$$

Identifying $f(x) = e^x / \left(1 + x^2\right)$ we obtain

$$u_1' = -\frac{x e^x e^x}{e^{2x} \left(1 + x^2\right)} = -\frac{x}{1 + x^2}$$

$$u_2' = \frac{e^x e^x}{e^{2x} \left(1 + x^2\right)} = \frac{1}{1 + x^2}.$$

Then $u_1 = -\frac{1}{2} \ln \left(1 + x^2\right)$, $u_2 = \tan^{-1} x$, and

$$y = c_1 e^x + c_2 x e^x - \frac{1}{2} e^x \ln \left(1 + x^2\right) + x e^x \tan^{-1} x$$

for $-\infty < x < \infty$.

18. The auxiliary equation is $m^2 + 10m + 25 = (m+5)^2 = 0$, so $y_c = c_1 e^{-5x} + c_2 x e^{-5x}$ and

$$W = \begin{vmatrix} e^{-5x} & x e^{-5x} \\ -5 e^{-5x} & -5x e^{-5x} + e^{-5x} \end{vmatrix} = e^{-10x}.$$

Identifying $f(x) = e^{-10x}/x^2$ we obtain

$$u_1' = -\frac{x e^{-5x} e^{-10x}}{x^2 e^{-10x}} = -\frac{e^{-5x}}{x}$$

$$u_2' = \frac{e^{-5x} e^{-10x}}{x^2 e^{-10x}} = \frac{e^{-5x}}{x^2}.$$

Then

$$u_1 = -\int_{x_0}^{x} \frac{e^{-5t}}{t}\, dt, \quad x_0 > 0$$

$$u_2 = \int_{x_0}^{x} \frac{e^{-5t}}{t^2}\, dt, \quad x_0 > 0$$

and

$$y = c_1 e^{-5x} + c_2 x e^{-5x} - e^{-5x} \int_{x_0}^{x} \frac{e^{-5t}}{t}\, dt + x e^{-5x} \int_{x_0}^{x} \frac{e^{-5t}}{t^2}\, dt$$

37

for $x > 0$.

21. The auxiliary equation is $m^3 + m = m(m^2 + 1) = 0$, so $y_c = c_1 + c_2 \cos x + c_3 \sin x$ and

$$W = \begin{vmatrix} 1 & \cos x & \sin x \\ 0 & -\sin x & \cos x \\ 0 & -\cos x & -\sin x \end{vmatrix} = 1.$$

Identifying $f(x) = \tan x$ we obtain

$$u_1' = W_1 = \begin{vmatrix} 0 & \cos x & \sin x \\ 0 & -\sin x & \cos x \\ \tan x & -\cos x & -\sin x \end{vmatrix} = \tan x$$

$$u_2' = W_2 = \begin{vmatrix} 1 & 0 & \sin x \\ 0 & 0 & \cos x \\ 0 & \tan x & -\sin x \end{vmatrix} = -\sin x$$

$$u_3' = W_3 = \begin{vmatrix} 1 & \cos x & 0 \\ 0 & -\sin x & 0 \\ 0 & -\cos x & \tan x \end{vmatrix} = -\sin x \tan x = \frac{\cos^2 x - 1}{\cos x} = \cos x - \sec x.$$

Then

$$u_1 = -\ln|\cos x|$$

$$u_2 = \cos x$$

$$u_3 = \sin x - \ln|\sec x + \tan x|$$

and

$$y = c_1 + c_2 \cos x + c_3 \sin x - \ln|\cos x| + \cos^2 x$$

$$+ \sin^2 x - \sin x \ln|\sec x + \tan x|$$

$$= c_4 + c_2 \cos x + c_3 \sin x - \ln|\cos x| - \sin x \ln|\sec x + \tan x|$$

for $-\infty < x < \infty$.

24. The auxiliary equation is $2m^3 - 6m^2 = 2m^2(m - 3) = 0$, so $y_c = c_1 + c_2 x + c_3 e^{3x}$ and

$$W = \begin{vmatrix} 1 & x & e^{3x} \\ 0 & 1 & 3e^{3x} \\ 0 & 0 & 9e^{3x} \end{vmatrix} = 9e^{3x}.$$

Identifying $f(x) = x^2/2$ we obtain

$$u_1' = \frac{1}{9e^{3x}} W_1 = \frac{1}{9e^{3x}} \begin{vmatrix} 0 & x & e^{3x} \\ 0 & 1 & 3e^{3x} \\ x^2/2 & 0 & 9e^{3x} \end{vmatrix} = \frac{\frac{3}{2}x^3 e^{3x} - \frac{1}{2}x^2 e^{3x}}{9e^{3x}} = \frac{1}{6}x^3 - \frac{1}{18}x^2$$

$$u_2' = \frac{1}{9e^{3x}} W_2 = \frac{1}{9e^{3x}} \begin{vmatrix} 1 & 0 & e^{3x} \\ 0 & 0 & 3e^{3x} \\ 0 & x^2/2 & 9e^{3x} \end{vmatrix} = \frac{-\frac{3}{2}x^2 e^{3x}}{9e^{3x}} = -\frac{1}{6}x^2$$

$$u_3' = \frac{1}{9e^{3x}} W_3 = \frac{1}{9e^{3x}} \begin{vmatrix} 1 & x & 0 \\ 0 & 1 & 0 \\ 0 & 0 & x^2/2 \end{vmatrix} = \frac{\frac{1}{2}x^2}{9e^{3x}} = \frac{1}{18}x^2 e^{-3x}.$$

Then

$$u_1 = \frac{1}{24}x^4 - \frac{1}{54}x^3$$

$$u_2 = -\frac{1}{18}x^3$$

$$u_3 = -\frac{1}{54}x^2 e^{-3x} - \frac{1}{81}xe^{-3x} - \frac{1}{243}e^{-3x}$$

and

$$y = c_1 + c_2 x + c_3 e^{3x} + \frac{1}{24}x^4 - \frac{1}{54}x^3 - \frac{1}{18}x^4 - \frac{1}{54}x^2 - \frac{1}{81}x - \frac{1}{243}$$

$$= c_4 + c_5 x + c_3 e^{3x} - \frac{1}{72}x^4 - \frac{1}{54}x^3 - \frac{1}{54}x^2$$

for $-\infty < x < \infty$.

27. The auxiliary equation is $m^2 + 2m - 8 = (m-2)(m+4) = 0$, so $y_c = c_1 e^{2x} + c_2 e^{-4x}$ and

$$W = \begin{vmatrix} e^{2x} & e^{-4x} \\ 2e^{2x} & -4e^{-4x} \end{vmatrix} = -6e^{-2x}.$$

Identifying $f(x) = 2e^{-2x} - e^{-x}$ we obtain

$$u_1' = \frac{1}{3}e^{-4x} - \frac{1}{6}e^{-3x}$$

$$u_2' = -\frac{1}{6}e^{3x} - \frac{1}{3}e^{2x}.$$

Then

$$u_1 = -\frac{1}{12}e^{-4x} + \frac{1}{18}e^{-3x}$$

$$u_2 = \frac{1}{18}e^{3x} - \frac{1}{6}e^{2x}.$$

Thus

$$y = c_1 e^{2x} + c_2 e^{-4x} - \frac{1}{12}e^{-2x} + \frac{1}{18}e^{-x} + \frac{1}{18}e^{-x} - \frac{1}{6}e^{-2x}$$

$$= c_1 e^{2x} + c_2 e^{-4x} - \frac{1}{4}e^{-2x} + \frac{1}{9}e^{-x}$$

and

$$y' = 2c_1 e^{2x} - 4c_2 e^{-4x} + \frac{1}{2}e^{-2x} - \frac{1}{9}e^{-x}.$$

The initial conditions imply

$$c_1 + c_2 - \frac{5}{36} = 1$$

$$2c_1 - 4c_2 + \frac{7}{18} = 0.$$

Thus $c_1 = 25/36$ and $c_2 = 4/9$, and

$$y = \frac{25}{36}e^{2x} + \frac{4}{9}e^{-4x} - \frac{1}{4}e^{-2x} + \frac{1}{9}e^{-x}.$$

30. Write the equation in the form

$$y'' + \frac{1}{x}y' + \frac{1}{x^2}y = \frac{\sec(\ln x)}{x^2}$$

and identify $f(x) = \sec(\ln x)/x^2$. From $y_1 = \cos(\ln x)$ and $y_2 = \sin(\ln x)$ we compute

$$W = \begin{vmatrix} \cos(\ln x) & \sin(\ln x) \\ -\dfrac{\sin(\ln x)}{x} & \dfrac{\cos(\ln x)}{x} \end{vmatrix} = \frac{1}{x}.$$

Now

$$u_1' = -\frac{\tan(\ln x)}{x} \quad \text{so} \quad u_1 = \ln|\cos(\ln x)|,$$

and

$$u_2' = \frac{1}{x} \quad \text{so} \quad u_2 = \ln x.$$

Thus, a particular solution is

$$y_p = \cos(\ln x)\ln|\cos(\ln x)| + (\ln x)\sin(\ln x).$$

Exercises 4.7

3. The auxiliary equation is $m^2 = 0$ so that $y = c_1 + c_2 \ln x$.

6. The auxiliary equation is $m^2 + 4m + 3 = (m+1)(m+3) = 0$ so that $y = c_1 x^{-1} + c_2 x^{-3}$.

9. The auxiliary equation is $25m^2 + 1 = 0$ so that $y = c_1 \cos\left(\frac{1}{5}\ln x\right) + c_2 \sin\left(\frac{1}{5}\ln x\right)$.

12. The auxiliary equation is $m^2 + 7m + 6 = (m+1)(m+6) = 0$ so that $y = c_1 x^{-1} + c_2 x^{-6}$.

15. The auxiliary equation is $3m^2 + 3m + 1 = 0$ so that $y = x^{-1/2}\left[c_1 \cos\left(\frac{\sqrt{3}}{6}\ln x\right) + c_2 \sin\left(\frac{\sqrt{3}}{6}\ln x\right)\right]$.

18. Assuming that $y = x^m$ and substituting into the differential equation we obtain

$$m(m-1)(m-2) + m - 1 = m^3 - 3m^2 + 3m - 1 = (m-1)^3 = 0.$$

Thus

$$y = c_1 x + c_2 x \ln x + c_3 x (\ln x)^2.$$

21. Assuming that $y = x^m$ and substituting into the differential equation we obtain

$$m(m-1)(m-2)(m-3) + 6m(m-1)(m-2) = m^4 - 7m^2 + 6m = m(m-1)(m-2)(m+3) = 0.$$

Thus

$$y = c_1 + c_2 x + c_3 x^2 + c_4 x^{-3}.$$

24. The auxiliary equation is $m^2 - 6m + 8 = (m-2)(m-4) = 0$, so that

$$y = c_1 x^2 + c_2 x^4 \quad\text{and}\quad y' = 2c_1 x + 4c_2 x^3.$$

The initial conditions imply

$$4c_1 + 16c_2 = 32$$

$$4c_1 + 32c_2 = 0.$$

Thus, $c_1 = 16$, $c_2 = -2$, and $y = 16x^2 - 2x^4$.

27. In this problem we use the substitution $t = -x$ since the initial conditions are on the interval $(-\infty, 0)$. Then

$$\frac{dy}{dt} = \frac{dy}{dx}\frac{dx}{dt} = -\frac{dy}{dx}$$

and

$$\frac{d^2y}{dt^2} = \frac{d}{dt}\left(\frac{dy}{dt}\right) = \frac{d}{dt}\left(-\frac{dy}{dx}\right) = -\frac{d}{dt}(y') = -\frac{dy'}{dx}\frac{dx}{dt} = -\frac{d^2y}{dx^2}\frac{dx}{dt} = \frac{d^2y}{dx^2},$$

so the differential equation and initial conditions become

$$4t^2 \frac{d^2y}{dt^2} + y = 0; \quad y(t)\Big|_{t=1} = 2, \quad y'(t)\Big|_{t=1} = -4.$$

The auxiliary equation is $4m^2 - 4m + 1 = (2m - 1)^2 = 0$, so that

$$y = c_1 t^{1/2} + c_2 t^{1/2} \ln t \quad \text{and} \quad y' = \frac{1}{2}c_1 t^{-1/2} + c_2 \left(t^{-1/2} + \frac{1}{2}t^{-1/2} \ln t \right).$$

The initial conditions imply $c_1 = 2$ and $1 + c_2 = -4$. Thus

$$y = 2t^{1/2} - 5t^{1/2} \ln t = 2(-x)^{1/2} - 5(-x)^{1/2} \ln(-x), \quad x < 0.$$

30. The auxiliary equation is $m^2 - 5m = m(m - 5) = 0$ so that $y_c = c_1 + c_2 x^5$ and

$$W(1, x^5) = \begin{vmatrix} 1 & x^5 \\ 0 & 5x^4 \end{vmatrix} = 5x^4.$$

Identifying $f(x) = x^3$ we obtain $u_1' = -\frac{1}{5}x^4$ and $u_2' = 1/5x$. Then $u_1 = -\frac{1}{25}x^5$, $u_2 = \frac{1}{5}\ln x$, and

$$y = c_1 + c_2 x^5 - \frac{1}{25}x^5 + \frac{1}{5}x^5 \ln x = c_1 + c_3 x^5 + \frac{1}{5}x^5 \ln x.$$

33. The auxiliary equation is $m^2 - 2m + 1 = (m - 1)^2 = 0$ so that $y_c = c_1 x + c_2 x \ln x$ and

$$W(x, x \ln x) = \begin{vmatrix} x & x \ln x \\ 1 & 1 + \ln x \end{vmatrix} = x.$$

Identifying $f(x) = 2/x$ we obtain $u_1' = -2 \ln x / x$ and $u_2' = 2/x$. Then $u_1 = -(\ln x)^2$, $u_2 = 2 \ln x$, and

$$y = c_1 x + c_2 x \ln x - x(\ln x)^2 + 2x(\ln x)^2$$

$$= c_1 x + c_2 x \ln x + x(\ln x)^2.$$

In Problems 36-39 we use the following results: When $x = e^t$ or $t = \ln x$, then

$$\frac{dy}{dx} = \frac{1}{x}\frac{dy}{dt} \quad \text{and} \quad \frac{d^2y}{dx^2} = \frac{1}{x^2}\left[\frac{d^2y}{dt^2} - \frac{dy}{dt} \right].$$

36. Substituting into the differential equation we obtain

$$\frac{d^2y}{dt^2} - 5\frac{dy}{dt} + 6y = 2t.$$

The auxiliary equation is $m^2 - 5m + 6 = (m - 2)(m - 3) = 0$ so that $y_c = c_1 e^{2t} + c_2 e^{3t}$. Using undetermined coefficients we try $y_p = At + B$. This leads to $(-5A + 6B) + 6At = 2t$, so that $A = 1/3$, $B = 5/18$, and

$$y = c_1 e^{2t} + c_2 e^{3t} + \frac{1}{3}t + \frac{5}{18} = c_1 x^2 + c_2 x^3 + \frac{1}{3}\ln x + \frac{5}{18}.$$

39. Substituting into the differential equation we obtain

$$\frac{d^2y}{dt^2} + 8\frac{dy}{dt} - 20y = 5e^{-3t}.$$

The auxiliary equation is $m^2 + 8m - 20 = (m+10)(m-2) = 0$ so that $y_c = c_1 e^{-10t} + c_2 e^{2t}$. Using undetermined coefficients we try $y_p = Ae^{-3t}$. This leads to $-35Ae^{-3t} = 5e^{-3t}$, so that $A = -1/7$ and

$$y = c_1 e^{-10t} + c_2 e^{2t} - \frac{1}{7}e^{-3t} = c_1 x^{-10} + c_2 x^2 - \frac{1}{7}x^{-3}.$$

Exercises 4.8

3. From $Dx = -y + t$ and $Dy = x - t$ we obtain $y = t - Dx$, $Dy = 1 - D^2x$, and $(D^2 + 1)x = 1 + t$. Then

$$x = c_1 \cos t + c_2 \sin t + 1 + t$$

and

$$y = c_1 \sin t - c_2 \cos t + t - 1.$$

6. From $(D+1)x + (D-1)y = 2$ and $3x + (D+2)y = -1$ we obtain $x = -\frac{1}{3} - \frac{1}{3}(D+2)y$, $Dx = -\frac{1}{3}(D^2 + 2D)y$, and $(D^2 + 5)y = -7$. Then

$$y = c_1 \cos \sqrt{5}\,t + c_2 \sin \sqrt{5}\,t - \frac{7}{5}$$

and

$$x = \left(-\frac{2}{3}c_1 - \frac{\sqrt{5}}{3}c_2\right)\cos \sqrt{5}\,t + \left(\frac{\sqrt{5}}{3}c_1 - \frac{2}{3}c_2\right)\sin \sqrt{5}\,t + \frac{3}{5}.$$

9. From $Dx + D^2y = e^{3t}$ and $(D+1)x + (D-1)y = 4e^{3t}$ we obtain $D(D^2 + 1)x = 34e^{3t}$ and $D(D^2 + 1)y = -8e^{3t}$. Then

$$y = c_1 + c_2 \sin t + c_3 \cos t - \frac{4}{15}e^{3t}$$

and

$$x = c_4 + c_5 \sin t + c_6 \cos t + \frac{17}{15}e^{3t}.$$

Substituting into $(D+1)x + (D-1)y = 4e^{3t}$ gives

$$(c_4 - c_1) + (c_5 - c_6 - c_3 - c_2)\sin t + (c_6 + c_5 + c_2 - c_3)\cos t = 0$$

so that $c_4 = c_1$, $c_5 = c_3$, $c_6 = -c_2$, and

$$x = c_1 - c_2 \cos t + c_3 \sin t + \frac{17}{15}e^{3t}.$$

43

12. From $(2D^2-D-1)x-(2D+1)y = 1$ and $(D-1)x+Dy = -1$ we obtain $(2D+1)(D-1)(D+1)x = -1$ and $(2D+1)(D+1)y = -2$. Then

$$x = c_1 e^{-t/2} + c_2 e^{-t} + c_3 e^t + 1$$

and

$$y = c_4 e^{-t/2} + c_5 e^{-t} - 2.$$

Substituting into $(D-1)x + Dy = -1$ gives

$$\left(-\frac{3}{2}c_1 - \frac{1}{2}c_4\right)e^{-t/2} + (-2c_2 - c_5)e^{-t} = 0$$

so that $c_4 = -3c_1$, $c_5 = -2c_2$, and

$$y = -3c_1 e^{-t/2} - 2c_2 e^{-t} - 2.$$

15. From $(D-1)x + (D^2 + 1)y = 1$ and $(D^2 - 1)x + (D+1)y = 2$ we obtain $D^2(D-1)(D+1)x = 1$ and $D^2(D-1)(D+1)y = 1$. Then

$$x = c_1 + c_2 t + c_3 e^t + c_4 e^{-t} - \frac{1}{2}t^2$$

and

$$y = c_5 + c_6 t + c_7 e^t + c_8 e^{-t} - \frac{1}{2}t^2.$$

Substituting into $(D-1)x + (D^2 + 1)y = 1$ gives

$$(c_2 - c_1 - 1 + c_5) + (c_6 - c_2 - 1)t + (2c_8 - 2c_4)e^{-t} + (2c_7)e^t = 1$$

so that $c_6 = c_2 + 1$, $c_8 = c_4$, $c_7 = 0$, $c_5 = c_1 - c_2 + 2$, and

$$y = (c_1 - c_2 + 2) + (c_2 + 1)t + c_4 e^{-t} - \frac{1}{2}t^2.$$

18. From $Dx + z = e^t$, $(D-1)x + Dy + Dz = 0$, and $x + 2y + Dz = e^t$ we obtain $z = -Dx + e^t$, $Dz = -D^2 x + e^t$, and the system $(-D^2 + D - 1)x + Dy = -e^t$ and $(-D^2 + 1)x + 2y = 0$. Then $y = \frac{1}{2}(D^2 - 1)x$, $Dy = \frac{1}{2}D(D^2 - 1)x$, and $(D-2)(D^2 + 1)x = -2e^t$ so that

$$x = c_1 e^{2t} + c_2 \cos t + c_3 \sin t + e^t,$$

$$y = \frac{3}{2}c_1 e^{2t} - c_2 \cos t - c_3 \sin t,$$

and

$$z = -2c_1 e^{2t} - c_3 \cos t + c_2 \sin t.$$

21. From $2Dx + (D-1)y = t$ and $Dx + Dy = t^2$ we obtain $(D+1)y = 2t^2 - t$. Then

$$y = c_1 e^{-t} + 2t^2 - 5t + 5$$

and $Dx = c_1 e^{-t} + t^2 - 4t + 5$ so that

$$x = -c_1 e^{-t} + c_2 + \frac{1}{3}t^3 - 2t^2 + 5t.$$

21. From $Dx - y = -1$ and $3x + (D-2)y = 0$ we obtain $x = -\frac{1}{3}(D-2)y$ so that $Dx = -\frac{1}{3}(D^2 - 2D)y$. Then $-\frac{1}{3}(D^2 - 2D)y = y - 1$ and $(D^2 - 2D + 3)y = 3$. Thus

$$y = e^t \left(c_1 \cos \sqrt{2}\, t + c_2 \sin \sqrt{2}\, t \right) + 1$$

and

$$x = \frac{1}{3} e^t \left[\left(c_1 - \sqrt{2}\, c_2\right) \cos \sqrt{2}\, t + \left(\sqrt{2}\, c_1 + c_2\right) \sin \sqrt{2}\, t \right] + \frac{2}{3}.$$

Using $x(0) = y(0) = 0$ we obtain

$$c_1 + 1 = 0$$

$$\frac{1}{3}\left(c_1 - \sqrt{2}\, c_2\right) + \frac{2}{3} = 0.$$

Thus $c_1 = -1$ and $c_2 = \sqrt{2}/2$. The solution of the initial value problem is

$$x = e^t \left(-\frac{2}{3} \cos \sqrt{2}\, t - \frac{\sqrt{2}}{6} \sin \sqrt{2}\, t \right) + \frac{2}{3}$$

$$y = e^t \left(-\cos \sqrt{2}\, t + \frac{\sqrt{2}}{2} \sin \sqrt{2}\, t \right) + 1.$$

Exercises 4.9

3. Let $u = y'$ so that $u' = y''$. The equation becomes $u' = -u - 1$ which is separable. Thus

$$\frac{du}{u^2 + 1} = -dx \implies \tan^{-1} u = -x + c_1 \implies y' = \tan(c_1 - x) \implies y = \ln|\cos(c_1 - x)| + c_2.$$

6. Let $u = y'$ so that $y'' = u \dfrac{du}{dy}$. The equation becomes $(y+1)u \dfrac{du}{dy} = u^2$. Separating variables we obtain

$$\frac{du}{u} = \frac{dy}{y + 1} \implies \ln|u| = \ln|y + 1| + \ln c_1 \implies u = c_1(y + 1)$$

$$\implies \frac{dy}{dx} = c_1(y + 1) \implies \frac{dy}{y + 1} = c_1 \, dx$$

$$\implies \ln|y + 1| = c_1 x + c_2 \implies y + 1 = c_3 e^{c_1 x}.$$

9. Let $u = y'$ so that $y'' = u\dfrac{du}{dy}$. The equation becomes $u\dfrac{du}{dy} + yu = 0$. Separating variables we obtain

$$du = -y\,dy \implies u = -\frac{1}{2}y^2 + c_1 \implies y' = -\frac{1}{2}y^2 + c_1.$$

When $x = 0$, $y = 1$ and $y' = -1$ so $-1 = -\frac{1}{2} + c_1$ and $c_1 = -\frac{1}{2}$. Then

$$\frac{dy}{dx} = -\frac{1}{2}y^2 - \frac{1}{2} \implies \frac{dy}{y^2 + 1} = -\frac{1}{2}\,dx \implies \tan^{-1}y = -\frac{1}{2}x + c_2$$

$$\implies y = \tan\left(-\frac{1}{2}x + c_2\right).$$

When $x = 0$, $y = 1$ so $1 = \tan c_2$ and $c_2 = \pi/4$. The solution of the initial-value problem is

$$y = \tan\left(\frac{\pi}{4} - \frac{1}{2}x\right), \qquad -\frac{\pi}{2} < x < \frac{3\pi}{2}.$$

12. Let $u = y'$ so that $u' = y''$. The equation becomes $u' - \dfrac{1}{x}u = u^2$, which is Bernoulli. Using the substitution $w = u^{-1}$ we obtain $\dfrac{dw}{dx} + \dfrac{1}{x}w = -1$. An integrating factor is x, so

$$\frac{d}{dx}[xw] = -x \implies w = -\frac{1}{2}x + \frac{1}{x}c \implies \frac{1}{u} = \frac{c_1 - x^2}{2x} \implies u = \frac{2x}{c_1 - x^2} \implies y = -\ln\left|c_1 - x^2\right| + c_2.$$

15. We look for a solution of the form

$$y(x) = y(0) + y'(0) + \frac{1}{2}y''(0) + \frac{1}{3!}y'''(0) + \frac{1}{4!}y^{(4)}(x) + \frac{1}{5!}y^{(5)}(x).$$

From $y''(x) = x^2 + y^2 - 2y'$ we compute

$$y'''(x) = 2x + 2yy' - 2y''$$

$$y^{(4)}(x) = 2 + 2(y')^2 + 2yy'' - 2y'''$$

$$y^{(5)}(x) = 6y'y'' + 2yy''' - 2y^{(4)}.$$

Using $y(0) = 1$ and $y'(0) = 1$ we find

$$y''(0) = -1, \quad y'''(0) = 4, \quad y^{(4)}(0) = -6, \quad y^{(5)}(0) = 14.$$

An approximate solution is

$$y(x) = 1 + x - \frac{1}{2}x^2 + \frac{2}{3}x^3 - \frac{1}{4}x^4 + \frac{7}{60}x^5.$$

18. Let $u = \dfrac{dx}{dt}$ so that $\dfrac{d^2x}{dt^2} = u\dfrac{du}{dx}$. The equation becomes $u\dfrac{du}{dx} = \dfrac{-k^2}{x^2}$. Separating variables we obtain

$$u\,du = -\frac{k^2}{x^2}\,dx \implies \frac{1}{2}u^2 = \frac{k^2}{x} + c \implies \frac{1}{2}v^2 = \frac{k^2}{x} + c.$$

When $t = 0$, $x = x_0$ and $v = 0$ so $0 = \dfrac{k^2}{x_0} + c$ and $c = -\dfrac{k^2}{x_0}$. Then

$$\frac{1}{2}v^2 = k^2 \left(\frac{1}{x} - \frac{1}{x_0} \right) \quad \text{and} \quad \frac{dx}{dt} = -k\sqrt{2}\sqrt{\frac{x_0 - x}{xx_0}}.$$

Separating variables we have

$$-\sqrt{\frac{xx_0}{x_0 - x}}\,dx = k\sqrt{2}\,dt \implies t = -\frac{1}{k}\sqrt{\frac{x_0}{2}} \int \sqrt{\frac{x}{x_0 - x}}\,dx.$$

Using *Mathematica* to integrate we obtain

$$t = -\frac{1}{k}\sqrt{\frac{x_0}{2}}\left[-\sqrt{x(x_0 - x)} - \frac{x_0}{2}\tan^{-1}\frac{(x_0 - 2x)}{2x}\sqrt{\frac{x}{x_0 - x}} \right]$$

$$= \frac{1}{k}\sqrt{\frac{x_0}{2}}\left[\sqrt{x(x_0 - x)} + \frac{x_0}{2}\tan^{-1}\frac{x_0 - 2x}{2\sqrt{x(x_0 - x)}} \right].$$

Chapter 4 Review Exercises

3. False; consider $f_1(x) = 0$ and $f_2(x) = x$. These are linearly dependent even though x is not a multiple of 0. The statement would be true if it read "Two functions $f_1(x)$ and $f_2(x)$ are linearly independent on an interval if *neither* is a constant multiple of the other."

6. True

9. $A + Bxe^x$

12. Identifying $P(x) = -2 - 2/x$ we have $\int P\,dx = -2x - 2\ln x$ and

$$y_2 = e^x \int \frac{e^{2x + \ln x^2}}{e^{2x}}\,dx = e^x \int x^2\,dx = \frac{1}{3}x^3 e^x.$$

15. From $m^3 + 10m^2 + 25m = 0$ we obtain $m = 0$, $m = -5$, and $m = -5$ so that

$$y = c_1 + c_2 e^{-5x} + c_3 x e^{-5x}.$$

18. From $2m^4 + 3m^3 + 2m^2 + 6m - 4 = 0$ we obtain $m = 1/2$, $m = -2$, and $m = \pm\sqrt{2}\,i$ so that

$$y = c_1 e^{x/2} + c_2 e^{-2x} + c_3 \cos \sqrt{2}\,x + c_4 \sin \sqrt{2}\,x.$$

21. Applying D^4 to the differential equation we obtain $D^4(D^2 - 3D + 5) = 0$. Then

$$y = \underbrace{e^{3x/2}\left(c_1 \cos \frac{\sqrt{11}}{2}x + c_2 \sin \frac{\sqrt{11}}{2}x \right)}_{y_c} + c_3 + c_4 x + c_5 x^2 + c_6 x^3$$

and $y_p = A + Bx + Cx^2 + Dx^3$. Substituting y_p into the differential equation yields

$$(5A - 3B + 2C) + (5B - 6C + 6D)x + (5C - 9D)x^2 + 5Dx^3 = -2x + 4x^3.$$

Equating coefficients gives $A = -222/625$, $B = 46/125$, $C = 36/25$, and $D = 4/5$. The general solution is

$$y = e^{3x/2}\left(c_1 \cos \frac{\sqrt{11}}{2}x + c_2 \sin \frac{\sqrt{11}}{2}x\right) - \frac{222}{625} + \frac{46}{125}x + \frac{36}{25}x^2 + \frac{4}{5}x^3.$$

24. Applying D to the differential equation we obtain $D(D^3 - D^2) = D^3(D - 1) = 0$. Then

$$y = \underbrace{c_1 + c_2x + c_3e^x}_{y_c} + c_4x^2$$

and $y_p = Ax^2$. Substituting y_p into the differential equation yields $-2A = 6$. Equating coefficients gives $A = -3$. The general solution is

$$y = c_1 + c_2x + c_3e^x - 3x^2.$$

27. Let $u = y'$ so that $u' = y''$. The equation becomes $u\dfrac{du}{dx} = 4x$. Separating variables we obtain

$$u\,du = 4x\,dx \implies \frac{1}{2}u^2 = 2x^2 + c_1 \implies u^2 = 4x^2 + c_2.$$

When $x = 1$, $y' = u = 2$, so $4 = 4 + c_2$ and $c_2 = 0$. Then

$$u^2 = 4x^2 \implies \frac{dy}{dx} = 2x \quad \text{or} \quad \frac{dy}{dx} = -2x$$

$$\implies y = x^2 + c_3 \quad \text{or} \quad y = -x^2 + c_4.$$

When $x = 1$, $y = 5$, so $5 = 1 + c_3$ and $5 = -1 + c_4$. Thus $c_3 = 4$ and $c_4 = 6$. We have $y = x^2 + 4$ and $y = -x^2 + 6$. Note however that when $y = -x^2 + 6$, $y' = -2x$ and $y'(1) = -2 \neq 2$. Thus, the solution of the initial-value problem is $y = x^2 + 4$.

30. The auxiliary equation is $m^2 - 1 = 0$, so $y_c = c_1e^x + c_2e^{-x}$ and

$$W = \begin{vmatrix} e^x & e^{-x} \\ e^x & -e^{-x} \end{vmatrix} = -2.$$

Identifying $f(x) = 2e^x/(e^x + e^{-x})$ we obtain

$$u_1' = \frac{1}{e^x + e^{-x}} = \frac{e^x}{1 + e^{2x}}$$

$$u_2' = -\frac{e^{2x}}{e^x + e^{-x}} = -\frac{e^{3x}}{1 + e^{2x}} = -e^x + \frac{e^x}{1 + e^{2x}}.$$

Then $u_1 = \tan^{-1} e^x$, $u_2 = -e^x + \tan^{-1} e^x$, and

$$y = c_1e^x + c_2e^{-x} + e^x \tan^{-1} e^x - 1 + e^{-x} \tan^{-1} e^x.$$

33. The auxiliary equation is $2m^3 - 13m^2 + 24m - 9 = (2m-1)(m-3)^2 = 0$ so that

$$y_c = c_1 e^{x/2} + c_2 e^{3x} + c_3 x e^{3x}.$$

A particular solution is $y_p = -4$ and the general solution is

$$y = c_1 e^{x/2} + c_2 e^{3x} + c_3 x e^{3x} - 4.$$

Setting $y(0) = -4$, $y'(0) = 0$, and $y''(0) = \frac{5}{2}$ we obtain

$$c_1 + c_2 - 4 = -4$$

$$\frac{1}{2}c_1 + 3c_2 + c_3 = 0$$

$$\frac{1}{4}c_1 + 9c_2 + 6c_3 = \frac{5}{2}.$$

Solving this system we find $c_1 = \frac{2}{5}$, $c_2 = -\frac{2}{5}$, and $c_3 = 1$. Thus

$$y = \frac{2}{5}e^{x/2} - \frac{2}{5}e^{3x} + x e^{3x} - 4.$$

36. From $(D-2)x - y = t - 2$ and $-3x + (D-4)y = -4t$ we obtain $(D-1)(D-5)x = 9 - 8t$. Then

$$x = c_1 e^t + c_2 e^{5t} - \frac{8}{5}t - \frac{3}{25}$$

and

$$y = (D-2)x - t + 2 = -c_1 e^t + 3c_2 e^{5t} + \frac{16}{25} + \frac{11}{25}t.$$

5 Modeling with Higher-Order Differential Equations

―――――― **Exercises 5.1** ――――――――――――――

3. From $\frac{3}{4}x'' + 72x = 0$, $x(0) = -1/4$, and $x'(0) = 0$ we obtain $x = -\frac{1}{4}\cos 4\sqrt{6}\,t$.

6. From $50x'' + 200x = 0$, $x(0) = 0$, and $x'(0) = -10$ we obtain $x = -5\sin 2t$ and $x' = -10\cos 2t$.

9. From $\frac{1}{4}x'' + x = 0$, $x(0) = 1/2$, and $x'(0) = 3/2$ we obtain

$$x = \frac{1}{2}\cos 2t + \frac{3}{4}\sin 2t = \frac{\sqrt{13}}{4}\sin(2t + 0.588).$$

12. From $x' + 9x = 0$, $x(0) = -1$, and $x'(0) = -\sqrt{3}$ we obtain

$$x = -\cos 3t - \frac{\sqrt{3}}{3}\sin 3t = \frac{2}{\sqrt{3}}\sin\left(37 + \frac{4\pi}{3}\right)$$

and $x' = 2\sqrt{3}\cos(3t + 4\pi/3)$. If $x' = 3$ then $t = -7\pi/18 + 2n\pi/3$ and $t = -\pi/2 + 2n\pi/3$ for $n = 1$, 2, 3, ...

18. (a) below   (b) from rest

21. From $\frac{1}{8}x'' + x' + 2x = 0$, $x(0) = -1$, and $x'(0) = 8$ we obtain $x = 4te^{-4t} - e^{-4t}$ and $x' = 8e^{-4t} - 16te^{-4t}$. If $x = 0$ then $t = 1/4$ second. If $x' = 0$ then $t = 1/2$ second and the extreme displacement is $x = e^{-2}$ feet.

24. (a) $x = \frac{1}{3}e^{-8t}\left(4e^{6t} - 1\right)$ is never zero; the extreme displacement is $x(0) = 1$ meter.

(b) $x = \frac{1}{3}e^{-8t}\left(5 - 2e^{6t}\right) = 0$ when $t = \frac{1}{6}\ln\frac{5}{2} \approx 0.153$ second; if $x' = \frac{4}{3}e^{-8t}\left(e^{6t} - 10\right) = 0$ then $t = \frac{1}{6}\ln 10 \approx 0.384$ second and the extreme displacement is $x = -0.232$ meter.

27. From $\frac{5}{16}x'' + \beta x' + 5x = 0$ we find that the roots of the auxiliary equation are $m = -\frac{8}{5}\beta \pm \frac{4}{5}\sqrt{4\beta^2 - 25}$.

(a) If $4\beta^2 - 25 > 0$ then $\beta > 5/2$.
(b) If $4\beta^2 - 25 = 0$ then $\beta = 5/2$.
(c) If $4\beta^2 - 25 < 0$ then $0 < \beta < 5/2$.

30. (a) If $x'' + 2x' + 5x = 12\cos 2t + 3\sin 2t$, $x(0) = -1$, and $x'(0) = 5$ then $x_c = e^{-t}(c_1\cos 2t + c_2\sin 2t)$ and $x_p = 3\sin 2t$ so that the equation of motion is

$$x = e^{-t}\cos 2t + 3\sin 2t.$$

(b)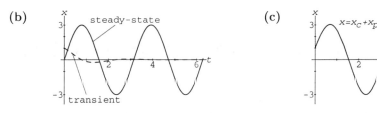

(c)

33. From $2x'' + 32x = 68e^{-2t}\cos 4t$, $x(0) = 0$, and $x'(0) = 0$ we obtain $x_c = c_1\cos 4t + c_2\sin 4t$ and $x_p = \frac{1}{2}e^{-2t}\cos 4t - 2e^{-2t}\sin 4t$ so that

$$x = -\frac{1}{2}\cos 4t + \frac{9}{4}\sin 4t + \frac{1}{2}e^{-2t}\cos 4t - 2e^{-2t}\sin 4t.$$

36. (a) From $100x'' + 1600x = 1600\sin 8t$, $x(0) = 0$, and $x'(0) = 0$ we obtain $x_c = c_1\cos 4t + c_2\sin 4t$ and $x_p = -\frac{1}{3}\sin 8t$ so that

$$x = \frac{2}{3}\sin 4t - \frac{1}{3}\sin 8t.$$

(b) If $x = \frac{1}{3}\sin 4t(2 - 2\cos 4t) = 0$ then $t = n\pi/4$ for $n = 0, 1, 2, \ldots$.

(c) If $x' = \frac{8}{3}\cos 4t - \frac{8}{3}\cos 8t = \frac{8}{3}(1 - \cos 4t)(1 + 2\cos 4t) = 0$ then $t = \pi/3 + n\pi/2$ and $t = \pi/6 + n\pi/2$ for $n = 0, 1, 2, \ldots$ at the extreme values. *Note:* There are many other values of t for which $x' = 0$.

(d) $x(\pi/6 + n\pi/2) = \sqrt{3}/2$ cm. and $x(\pi/3 + n\pi/2) = -\sqrt{3}/2$ cm.

(e)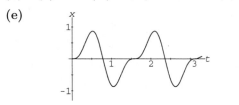

39. (a) From $x'' + \omega^2 x = F_0\cos\gamma t$, $x(0) = 0$, and $x'(0) = 0$ we obtain $x_c = c_1\cos\omega t + c_2\sin\omega t$ and $x_p = (F_0\cos\gamma t)/(\omega^2 - \gamma^2)$ so that

$$x = -\frac{F_0}{\omega^2 - \gamma^2}\cos\omega t + \frac{F_0}{\omega^2 - \gamma^2}\cos\gamma t.$$

(b) $\displaystyle\lim_{\gamma\to\omega}\frac{F_0}{\omega^2 - \gamma^2}(\cos\gamma t - \cos\omega t) = \lim_{\gamma\to\omega}\frac{-F_0 t\sin\gamma t}{-2\gamma} = \frac{F_0}{2\omega}t\sin\omega t.$

45. Solving $\frac{1}{20}q'' + 2q' + 100q = 0$ we obtain $q(t) = e^{-20t}(c_1\cos 40t + c_2\sin 40t)$. The initial conditions $q(0) = 5$ and $q'(0) = 0$ imply $c_1 = 5$ and $c_2 = 5/2$. Thus

$$q(t) = e^{-20t}\left(5\cos 40t + \frac{5}{2}\sin 40t\right) \approx \sqrt{25 + 25/4}\,e^{-20t}\sin(40t + 1.1071)$$

and $q(0.01) \approx 4.5676$ coulombs. The charge is zero for the first time when $40t + 0.4636 = \pi$ or $t \approx 0.0509$ second.

48. Solving $q'' + 100q' + 2500q = 30$ we obtain $q(t) = c_1 e^{-50t} + c_2 t e^{-50t} + 0.012$. The initial conditions $q(0) = 0$ and $q'(0) = 2$ imply $c_1 = -0.012$ and $c_2 = 1.4$. Thus

$$q(t) = -0.012e^{-50t} + 1.4te^{-50t} + 0.012 \quad \text{and} \quad i(t) = 2e^{-50t} - 70te^{-50t}.$$

Solving $i(t) = 0$ we see that the maximum charge occurs when $t = 1/35$ and $q(1/35) \approx 0.01871$.

51. The differential equation is $\frac{1}{2}q'' + 20q' + 1000q = 100\sin t$. To use Example 11 in the text we identify $E_0 = 100$ and $\gamma = 60$. Then

$$X = L\gamma - \frac{1}{c\gamma} = \frac{1}{2}(60) - \frac{1}{0.001(60)} \approx 13.3333,$$

$$Z = \sqrt{X^2 + R^2} = \sqrt{X^2 + 400} \approx 24.0370,$$

and

$$\frac{E_0}{Z} = \frac{100}{Z} \approx 4.1603.$$

From Problem 50, then

$$i_p(t) \approx 4.1603(60t + \phi)$$

where $\sin\phi = -X/Z$ and $\cos\phi = R/Z$. Thus $\tan\phi = -X/R \approx -0.6667$ and ϕ is a fourth quadrant angle. Now $\phi \approx -0.5880$ and

$$i_p(t) \approx 4.1603(60t - 0.5880).$$

54. By Problem 50 the amplitude of the steady-state current is E_0/Z, where $Z = \sqrt{X^2 + R^2}$ and $X = L\gamma - 1/C\gamma$. Since E_0 is constant the amplitude will be a maximum when Z is a minimum. Since R is constant, Z will be a minimum when $X = 0$. Solving $L\gamma - 1/C\gamma = 0$ for γ we obtain $\gamma = 1/\sqrt{LC}$. The maximum amplitude will be E_0/R.

57. In an L-C series circuit there is no resistor, so the differential equation is

$$L\frac{d^2q}{dt^2} + \frac{1}{C}q = E(t).$$

Then $q(t) = c_1 \cos\left(t/\sqrt{LC}\right) + c_2 \sin\left(t/\sqrt{LC}\right) + q_p(t)$ where $q_p(t) = A\sin\gamma t + B\cos\gamma t$. Substituting $q_p(t)$ into the differential equation we find

$$\left(\frac{1}{C} - L\gamma^2\right)A\sin\gamma t + \left(\frac{1}{C} - L\gamma^2\right)B\cos\gamma t = E_0\cos\gamma t.$$

Equating coefficients we obtain $A = 0$ and $B = E_0 C/(1 - LC\gamma^2)$. Thus, the charge is

$$q(t) = c_1 \cos\frac{1}{\sqrt{LC}}t + c_2 \sin\frac{1}{\sqrt{LC}}t + \frac{E_0 C}{1 - LC\gamma^2}\cos\gamma t.$$

The initial conditions $q(0) = q_0$ and $q'(0) = i_0$ imply $c_1 = q_0 - E_0C/(1 - LC\gamma^2)$ and $c_2 = i_0\sqrt{LC}$. The current is

$$i(t) = -\frac{c_1}{\sqrt{LC}}\sin\frac{1}{\sqrt{LC}}t + \frac{c_2}{\sqrt{LC}}\cos\frac{1}{\sqrt{LC}}t - \frac{E_0C\gamma}{1 - LC\gamma^2}\sin\gamma t$$

$$= i_0\cos\frac{1}{\sqrt{LC}}t - \frac{1}{\sqrt{LC}}\left(q_0 - \frac{E_0C}{1 - LC\gamma^2}\right)\sin\frac{1}{\sqrt{LC}}t - \frac{E_0C\gamma}{1 - LC\gamma^2}\sin\gamma t.$$

Exercises 5.2

3. (a) The general solution is

$$y(x) = c_1 + c_2x + c_3x^2 + c_4x^3 + \frac{w_0}{24EI}x^4.$$

The boundary conditions are $y(0) = 0$, $y'(0) = 0$, $y(L) = 0$, $y''(L) = 0$. The first two conditions give $c_1 = 0$ and $c_2 = 0$. The conditions at $x = L$ give the system

$$c_3L^2 + c_4L^3 + \frac{w_0}{24EI}L^4 = 0$$

$$2c_3 + 6c_4L + \frac{w_0}{2EI}L^2 = 0.$$

Solving, we obtain $c_3 = w_0L^2/16EI$ and $c_4 = -5w_0L/48EI$. The deflection is

$$y(x) = \frac{w_0}{48EI}(3L^2x^2 - 5Lx^3 + 2x^4).$$

(b)

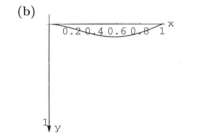

6. (a) $y_{max} = y(L/2) = \dfrac{5w_0L^4}{384EI}$

(b) The maximum deflection of the beam in Example 1 is $y(L/2) = (w_0/24EI)L^4/16 = w_0L^4/384EI$, which is $1/5$ of the maximum displacement of the beam in Problem 2.

9. For $\lambda \le 0$ the only solution of the boundary-value problem is $y = 0$. For $\lambda > 0$ we have

$$y = c_1\cos\sqrt{\lambda}\,x + c_2\sin\sqrt{\lambda}\,x.$$

Now $y(0) = 0$ implies $c_1 = 0$, so

$$y(\pi) = c_2 \sin \sqrt{\lambda}\, \pi = 0$$

gives

$$\sqrt{\lambda}\, \pi = n\pi \quad \text{or} \quad \lambda = n^2, \ n = 1, 2, 3, \ldots.$$

The eigenvalues n^2 correspond to the eigenfunctions $\sin nx$ for $n = 1, 2, 3, \ldots$.

12. For $\lambda \leq 0$ the only solution of the boundary-value problem is $y = 0$. For $\lambda > 0$ we have

$$y = c_1 \cos \sqrt{\lambda}\, x + c_2 \sin \sqrt{\lambda}\, x.$$

Now $y(0) = 0$ implies $c_1 = 0$, so

$$y'\left(\frac{\pi}{2}\right) = c_2 \sqrt{\lambda} \cos \sqrt{\lambda} \frac{\pi}{2} = 0$$

gives

$$\sqrt{\lambda} \frac{\pi}{2} = \frac{(2n-1)\pi}{2} \quad \text{or} \quad \lambda = (2n-1)^2, \ n = 1, 2, 3, \ldots.$$

The eigenvalues $(2n-1)^2$ correspond to the eigenfunctions $\sin(2n-1)x$.

15. The auxiliary equation has solutions

$$m = \frac{1}{2}\left(-2 \pm \sqrt{4 - 4(\lambda+1)}\right) = -1 \pm \sqrt{-\lambda}.$$

For $\lambda < 0$ we have

$$y = e^{-x}\left(c_1 \cosh \sqrt{-\lambda}\, x + c_2 \sinh \sqrt{-\lambda}\, x\right).$$

The boundary conditions imply

$$y(0) = c_1 = 0$$

$$y(5) = c_2 e^{-5} \sinh 5\sqrt{-\lambda} = 0$$

so $c_1 = c_2 = 0$ and the only solution of the boundary-value problem is $y = 0$.

For $\lambda = 0$ we have

$$y = c_1 e^{-x} + c_2 x e^{-x}$$

and the only solution of the boundary-value problem is $y = 0$.

For $\lambda > 0$ we have

$$y = e^{-x}\left(c_1 \cos \sqrt{\lambda}\, x + c_2 \sin \sqrt{\lambda}\, x\right).$$

Now $y(0) = 0$ implies $c_1 = 0$, so

$$y(5) = c_2 e^{-5} \sin 5\sqrt{\lambda} = 0$$

gives

$$5\sqrt{\lambda} = n\pi \quad \text{or} \quad \lambda = \frac{n^2\pi^2}{25}, \ n = 1, 2, 3, \ldots.$$

The eigenvalues $n^2\pi^2/25$ correspond to the eigenfunctions $e^{-x}\sin\dfrac{n\pi}{5}x$ for $n = 1, 2, 3, \ldots$.

18. For $\lambda = 0$ the only solution of the boundary-value problem is $y = 0$. For $\lambda \neq 0$ we have

$$y = c_1\cos\lambda x + c_2\sin\lambda x.$$

Now $y(0) = 0$ implies $c_1 = 0$, so

$$y'(3\pi) = c_2\lambda\cos 3\pi\lambda = 0$$

gives

$$3\pi\lambda = \frac{(2n-1)\pi}{2} \quad\text{or}\quad \lambda = \frac{2n-1}{6}, \quad n = 1, 2, 3, \ldots.$$

The eigenvalues $(2n-1)/6$ correspond to the eigenfunctions $\sin\dfrac{2n-1}{6}x$ for $n = 1, 2, 3, \ldots$.

21. For $\lambda = 0$ the general solution is $y = c_1 + c_2\ln x$. Now $y' = c_2/x$, so $y'(1) = c_2 = 0$ and $y = c_1$. Since $y'(e^2) = 0$ for any c_1 we see that $y(x) = 1$ is an eigenfunction corresponding to the eigenvalue $\lambda = 0$.

For $\lambda < 0$, $y = c_1 x^{-\sqrt{-\lambda}} + c_2 x^{\sqrt{-\lambda}}$. The initial conditions imply $c_1 = c_2 = 0$, so $y(x) = 0$.

For $\lambda > 0$, $y = c_1\cos(\sqrt{\lambda}\ln x) + c_2\sin(\sqrt{\lambda}\ln x)$. Now

$$y' = -c_1\frac{\sqrt{\lambda}}{x}\sin(\sqrt{\lambda}\ln x) + c_2\frac{\sqrt{\lambda}}{x}\cos(\sqrt{\lambda}\ln x),$$

and $y'(1) = c_2\sqrt{\lambda} - 0$ implies $c_2 = 0$. Finally, $y'(e^2) = -(c_1\sqrt{\lambda}/e^2)\sin(2\sqrt{\lambda}) = 0$ implies $\lambda = n^2\pi^2/4$ for $n = 1, 2, 3, \ldots$. The corresponding eigenfunctions are

$$y = \cos\left(\frac{n\pi}{2}\ln x\right).$$

24. (a) Since $\lambda_n = x_n^2$, there are no new eigenvalues when $x_n < 0$. For $\lambda = 0$, the differential equation $y'' = 0$ has general solution $y = c_1 x + c_2$. The boundary conditions imply $c_1 = c_2 = 0$, so $y = 0$.

(b) $\lambda_1 = 4.1159$, $\lambda_2 = 24.1393$, $\lambda_3 = 63.6591$, $\lambda_4 = 122.8892$.

27. The auxiliary equation is $m^2 + m = m(m+1) = 0$ so that $u(r) = c_1 r^{-1} + c_2$. The boundary conditions $u(a) = u_0$ and $u(b) = u_1$ yield the system $c_1 a^{-1} + c_2 = u_0$, $c_1 b^{-1} + c_2 = u_1$. Solving gives

$$c_1 = \left(\frac{u_0 - u_1}{b - a}\right)ab \quad\text{and}\quad c_2 = \frac{u_1 b - u_0 a}{b - a}.$$

Thus

$$u(r) = \left(\frac{u_0 - u_1}{b - a}\right)\frac{ab}{r} + \frac{u_1 b - u_0 a}{b - a}.$$

3. The period corresponding to $x(0) = 1$, $x'(0) = 1$ is approximately 5.8. The second initial-value problem does not have a periodic solution.

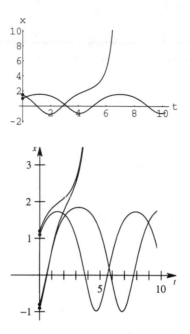

6. From the graphs we see that the interval is approximately $(-0.8, 1.1)$.

9. (a) This is a damped hard spring, so all solutions should be oscillatory with $x \to 0$ as $t \to \infty$.

(b)

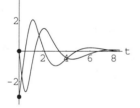

12. (a)

k = -0. 000471   k = -0. 000472

The system appears to be oscillatory for $-0.000471 \le k_1 < 0$ and nonoscillatory for $k_1 \le -0.000472$.

(b)

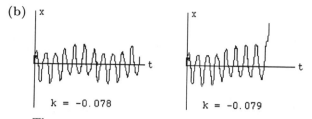

k = -0.078   k = -0.079

The system appears to be oscillatory for $-0.077 \leq k_1 < 0$ and nonoscillatory for $k_1 \leq 0.078$.

15. (a) Let (x, y) be the coordinates of S_2 on the curve C. The slope at (x, y) is then

$$dy/dx = (v_1 t - y)/(0 - x) = (y - v_1 t)/x \quad \text{or} \quad xy' - y = -v_1 t.$$

(b) Differentiating with respect to x gives

$$xy'' + y' - y' = -v_1 \frac{dt}{dx}$$

$$xy'' = -v_1 \frac{dt}{ds} \frac{ds}{dx}$$

$$xy'' = -v_1 \frac{1}{v_2}(-\sqrt{1 + (y')^2})$$

$$xy'' = r\sqrt{1 + (y')^2}.$$

Letting $u = y'$ and separating variables, we obtain

$$x \frac{du}{dx} = r\sqrt{1 + u^2}$$

$$\frac{du}{\sqrt{1 + u^2}} = \frac{r}{x} dx$$

$$\sinh^{-1} u = r \ln x + \ln c = \ln(cx^r)$$

$$u = \sinh(\ln cx^r)$$

$$\frac{dy}{dx} = \frac{1}{2}\left(cx^r - \frac{1}{cx^r}\right).$$

At $t = 0$, $dy/dx = 0$ and $x = a$, so $0 = ca^r - 1/ca^r$. Thus $c = 1/a^r$ and

$$\frac{dy}{dx} = \frac{1}{2}\left[\left(\frac{x}{a}\right)^r - \left(\frac{a}{x}\right)^r\right] = \frac{1}{2}\left[\left(\frac{x}{a}\right)^r - \left(\frac{x}{a}\right)^{-r}\right].$$

If $r > 1$ or $r < 1$, integrating gives

$$y = \frac{a}{2}\left[\frac{1}{1+r}\left(\frac{x}{a}\right)^{1+r} - \frac{1}{1-r}\left(\frac{x}{a}\right)^{1-r}\right] + c_1.$$

When $t = 0$, $y = 0$ and $x = a$, so $0 = (a/2)[1/(1+r) - 1/(1-r)] + c_1$. Thus $c_1 = ar/(1-r^2)$
and

$$y = \frac{a}{2}\left[\frac{1}{1+r}\left(\frac{x}{a}\right)^{1+r} - \frac{1}{1-r}\left(\frac{x}{a}\right)^{1-r}\right] + \frac{ar}{1-r^2}.$$

(c) If $r > 1$, $v_1 > v_2$ and $y \to \infty$ as $x \to 0^+$. In other words, S_2 always lags behind S_1. If $r < 1$,
$v_1 < v_2$ and $y = ar/(1-r^2)$ when $x = 0$. In other words, when the submarine's speed is greater
than the ship's, their paths will intersect at the point $(0, ar/(1-r^2))$.

If $r = 1$, integration gives

$$y = \frac{1}{2}\left[\frac{x^2}{2a} - \frac{1}{a}\ln x\right] + c_2.$$

When $t = 0$, $y = 0$ and $x = a$, so $0 = (1/2)[a/2 - (1/a)\ln a] + c_2$. Thus $c_2 = -(1/2)[a/2 - (1/a)\ln a]$ and

$$y = \frac{1}{2}\left[\frac{x^2}{2a} - \frac{1}{a}\ln x\right] - \frac{1}{2}\left[\frac{a}{2} - \frac{1}{a}\ln a\right] = \frac{1}{2}\left[\frac{1}{2a}(x^2 - a^2) + \frac{1}{a}\ln\frac{a}{x}\right].$$

Since $y \to \infty$ as $x \to 0^+$, S_2 will never catch up with S_1.

Chapter 5 Review Exercises

3. $5/4$ m., since $x = -\cos 4t + \frac{3}{4}\sin 4t$.

6. False

9. $9/2$, since $x = c_1 \cos \sqrt{2k}\, t + c_2 \sin \sqrt{2k}\, t$.

12. From $x'' + \beta x' + 64x = 0$ we see that oscillatory motion results if $\beta^2 - 256 < 0$ or $0 \le |\beta| < 16$.

15. Writing $\frac{1}{8}x'' + \frac{8}{3}x = \cos \gamma t + \sin \gamma t$ in the form $x'' + \frac{64}{3}x = 8\cos \gamma t + 8\sin \gamma t$ we identify $\lambda = 0$ and
$\omega^2 = 64/3$. The system is in a state of pure resonance when $\gamma = \omega = \sqrt{64/3} = 8/\sqrt{3}$.

18. (a) Let k be the effective spring constant and x_1 and x_2 the elongation of springs k_1 and k_2. The
restoring forces satisfy $k_1 x_1 = k_2 x_2$ so $x_2 = (k_1/k_2)x_1$. From $k(x_1 + x_2) = k_1 x_1$ we have

$$k\left(x_1 + \frac{k_1}{k_2}x_2\right) = k_1 x_1$$

$$k\left(\frac{k_2 + k_1}{k_2}\right) = k_1$$

$$k = \frac{k_1 k_2}{k_1 + k_2}$$

$$\frac{1}{k} = \frac{1}{k_1} + \frac{1}{k_2}.$$

(b) From $k_1 = 2W$ and $k_2 = 4W$ we find $1/k = 1/2W + 1/4W = 3/4W$. Then $k = 4W/3 = 4mg/3$. The differential equation $mx'' + kx = 0$ then becomes $x'' + (4g/3)x = 0$. The solution is

$$x(t) = c_1 \cos 2\sqrt{\frac{g}{3}} t + c_2 \sin 2\sqrt{\frac{g}{3}} t.$$

The initial conditions $x(0) = 1$ and $x'(0) = 2/3$ imply $c_1 = 1$ and $c_2 = 1/\sqrt{3g}$.

(c) To compute the maximum speed of the weight we compute

$$x'(t) = 2\sqrt{\frac{g}{3}} \sin 2\sqrt{\frac{g}{3}} t + \frac{2}{3} \cos 2\sqrt{\frac{g}{3}} t \quad \text{and} \quad |x'(t)| = \sqrt{4\frac{g}{3} + \frac{4}{9}} = \frac{2}{3}\sqrt{3g + 1}.$$

21. For $\lambda > 0$ the general solution is $y = c_1 \cos \sqrt{\lambda}\, x + c_2 \sin \sqrt{\lambda}\, x$. Now $y(0) = c_1$ and $y(2\pi) = c_1 \cos 2\pi\sqrt{\lambda} + c_2 \sin 2\pi\sqrt{\lambda}$, so the condition $y(0) = y(2\pi)$ implies

$$c_1 = c_1 \cos 2\pi\sqrt{\lambda} + c_2 \sin 2\pi\sqrt{\lambda}$$

which is true when $\sqrt{\lambda} = n$ or $\lambda = n^2$ for $n = 1, 2, 3, \ldots$. Since $y' = -\sqrt{\lambda}\, c_1 \sin \sqrt{\lambda}\, x + \sqrt{\lambda}\, c_2 \cos \sqrt{\lambda}\, x = -nc_1 \sin nx + nc_2 \cos nx$, we see that $y'(0) = nc_2 = y'(2\pi)$ for $n = 1, 2, 3, \ldots$. Thus, the eigenvalues are n^2 for $n = 1, 2, 3, \ldots$, with corresponding eigenfunctions $\cos nx$ and $\sin nx$. When $\lambda = 0$, the general solution is $y = c_1 x + c_2$ and the corresponding eigenfunction is $y = 1$. For $\lambda < 0$ the general solution is $y = c_1 \cosh \sqrt{-\lambda}\, x + c_2 \sinh \sqrt{-\lambda}\, x$. In this case $y(0) = c_1$ and $y(2\pi) = c_1 \cosh 2\pi\sqrt{-\lambda} + c_2 \sinh 2\pi\sqrt{-\lambda}$, so $y(0) = y(2\pi)$ can only be valid for $\lambda = 0$. Thus, there are no eigenvalues corresponding to $\lambda < 0$.

6 Series Solutions of Linear Equations

3. $\displaystyle\lim_{n\to\infty}\left|\frac{a_{n+1}}{a_n}\right| = \lim_{n\to\infty}\left|\frac{2^{n+1}x^{n+1}/(n+1)}{2^n x^n/n}\right| = \lim_{n\to\infty}\frac{2n}{n+1}|x| = 2|x|$

The series is absolutely convergent for $2|x| < 1$ or $|x| < 1/2$. At $x = -1/2$, the series $\displaystyle\sum_{k=1}^{\infty}\frac{(-1)^k}{k}$

converges by the alternating series test. At $x = 1/2$, the series $\displaystyle\sum_{k=1}^{\infty}\frac{1}{k}$ is the harmonic series which

diverges. Thus, the given series converges on $[-1/2, 1/2)$.

6. $\displaystyle\lim_{n\to\infty}\left|\frac{a_{n+1}}{a_n}\right| = \lim_{n\to\infty}\left|\frac{(x+7)^{n+1}/\sqrt{n+1}}{(x+7)^n\sqrt{n}}\right| = \lim_{n\to\infty}\sqrt{\frac{n}{n+1}}\,|x+7| = |x+7|$

The series is absolutely convergent for $|x+7| < 1$ or on $(-8, 6)$. At $x = -8$, the series $\displaystyle\sum_{n=1}^{\infty}\frac{(-1)^n}{\sqrt{n}}$

converges by the alternating series test. At $x = -6$, the series $\displaystyle\sum_{n=1}^{\infty}\frac{1}{\sqrt{n}}$ is a divergent p-series. Thus,

the given series converges on $[-8, -6)$.

9. $\displaystyle\lim_{n\to\infty}\left|\frac{a_{n+1}}{a_n}\right| = \lim_{n\to\infty}\left|\frac{(n+1)!2^{n+1}x^{n+1}}{n!2^n x^n}\right| = \lim_{n\to\infty}2(n+1)|x| = \infty, \quad x \neq 0$

The series converges only at $x = 0$.

12. $e^{-x}\cos x = \left(1 - x + \dfrac{x^2}{2} - \dfrac{x^3}{6} + \dfrac{x^4}{24} - \cdots\right)\left(1 - \dfrac{x^2}{2} + \dfrac{x^4}{24} - \cdots\right) = 1 - x + \dfrac{x^3}{3} - \dfrac{x^4}{6} + \cdots$

15. Separating variables we obtain

$$\frac{dy}{y} = -dx \implies \ln|y| = -x + c \implies y = c_1 e^{-x}.$$

Substituting $y = \sum_{n=0}^{\infty} c_n x^n$ into the differential equation leads to

$$y' + y = \underbrace{\sum_{n=1}^{\infty} nc_n x^{n-1}}_{k=n-1} + \underbrace{\sum_{n=0}^{\infty} c_n x^n}_{k=n} = \sum_{k=0}^{\infty}(k+1)c_{k+1}x^k + \sum_{k=0}^{\infty} c_k x^k = \sum_{k=0}^{\infty}[(k+1)c_{k+1} + c_k]x^k = 0.$$

Thus

$$(k+1)c_{k+1} + c_k = 0$$

and

$$c_{k+1} = -\frac{1}{k+1}c_k, \quad k = 0, 1, 2, \dots.$$

Iterating we find

$$c_1 = -c_0$$

$$c_2 = -\frac{1}{2}c_1 = \frac{1}{2}c_0$$

$$c_3 = -\frac{1}{3}c_2 = -\frac{1}{6}c_0$$

$$c_4 = -\frac{1}{4}c_3 = \frac{1}{24}c_0$$

and so on. Therefore

$$y = c_0 - c_0 x + \frac{1}{2}c_0 x^2 - \frac{1}{6}c_0 x^3 + \frac{1}{24}c_0 x^4 - \cdots = c_0\left[1 - x + \frac{1}{2}x^2 - \frac{1}{6}x^3 + \frac{1}{24}x^4 - \cdots\right]$$

$$= c_0 \sum_{n=0}^{\infty} \frac{1}{n!}(-x)^n = c_0 e^{-x}.$$

18. Separating variables we obtain

$$\frac{dy}{y} = -x^3 dx \implies \ln|y| = -\frac{1}{4}x^4 + c \implies y = c_1 e^{-x^4/4}.$$

Substituting $y = \sum_{n=0}^{\infty} c_n x^n$ into the differential equation leads to

$$y' + x^3 y = \underbrace{\sum_{n=1}^{\infty} n c_n x^{n-1}}_{k=n-4} + \underbrace{\sum_{n=0}^{\infty} c_n x^{n+3}}_{k=n} = \sum_{k=-3}^{\infty}(k+4)c_{k+4}x^{k+3} - \sum_{k=0}^{\infty} c_k x^{k+3}$$

$$= c_1 + 2c_2 x + 3c_3 x^2 + \sum_{k=0}^{\infty}[(k+4)c_{k+4} + c_k]x^{k+2} = 0.$$

Thus

$$c_1 = c_2 = c_3 = 0,$$

$$(k+4)c_{k+4} + c_k = 0,$$

and

$$c_{k+4} = -\frac{1}{k+4}c_k, \quad k = 0, 1, 2, \ldots.$$

Iterating we find

$$c_4 = -\frac{1}{4}c_0$$

$$c_5 = c_6 = c_7 = 0$$

$$c_8 = -\frac{1}{8}c_4 = \frac{1}{2}\cdot\frac{1}{4^2}c_0$$

$$c_9 = c_{10} = c_{11} = 0$$

$$c_{12} = -\frac{1}{12}c_8 = -\frac{1}{2\cdot3}\cdot\frac{1}{4^3}c_0$$

and so on. Therefore

$$y = c_0 - \frac{1}{4}c_0 x^4 + \frac{1}{2} \cdot \frac{1}{4^2}c_0 x^8 - \frac{1}{2 \cdot 3} \cdot \frac{1}{4^3}c_0 x^{12} + \cdots$$

$$= c_0 \left[1 - \frac{x^4}{4} + \frac{1}{2}\left(\frac{x^4}{4}\right)^2 - \frac{1}{2 \cdot 3}\left(\frac{x^4}{4}\right)^3 + \cdots \right] = c_0 \sum_{n=0}^{\infty} \frac{1}{n!}\left(\frac{-x^4}{4}\right)^n = c_0 e^{-x^4/4}.$$

21. The auxiliary equation is $m^2 + 1 = 0$, so $y = c_1 \cos x + c_2 \sin x$. Substituting $y = \sum_{n=0}^{\infty} c_n x^n$ into the differential equation leads to

$$y'' + y = \underbrace{\sum_{n=2}^{\infty} n(n-1)c_n x^{n-2}}_{k=n-2} + \underbrace{\sum_{n=0}^{\infty} c_n x^n}_{k=n} = \sum_{k=0}^{\infty}(k+2)(k+1)c_{k+2}x^k + \sum_{k=0}^{\infty} c_k x^k$$

$$= \sum_{k=0}^{\infty}[(k+2)(k+1)c_{k+2} + c_k]x^k = 0.$$

Thus

$$(k+2)(k+1)c_{k+2} + c_k = 0$$

and

$$c_{k+2} = -\frac{1}{(k+2)(k+1)}c_k, \quad k = 0, 1, 2, \ldots .$$

Iterating we find

$$c_2 = -\frac{1}{2}c_0$$

$$c_3 = -\frac{1}{3 \cdot 2}c_1$$

$$c_4 = -\frac{1}{4 \cdot 3}c_2 = \frac{1}{4 \cdot 3 \cdot 2}c_0$$

$$c_5 = -\frac{1}{5 \cdot 4}c_3 = \frac{1}{5 \cdot 4 \cdot 3 \cdot 2}c_1$$

$$c_6 = -\frac{1}{6 \cdot 5}c_4 = -\frac{1}{6!}c_0$$

$$c_7 = -\frac{1}{7 \cdot 6}c_5 = -\frac{1}{7!}c_1$$

and so on. Therefore

$$y = c_0 + c_1 x - \frac{1}{2}c_0 x^2 - \frac{1}{3!}c_1 x^3 + \frac{1}{4!}c_0 x^4 + \frac{1}{5!}c_1 x^5 - \cdots$$

$$= c_0 \left[1 - \frac{1}{2}x^2 + \frac{1}{4!}x^4 - \cdots \right] + c_1 \left[1 - \frac{1}{3!}x^3 + \frac{1}{5!}x^5 - \cdots \right]$$

$$= c_0 \sum_{n=0}^{\infty} \frac{(-1)^n x^{2n}}{(2n)!} + c_1 \sum_{n=0}^{\infty} \frac{(-1)^n x^{2n+1}}{(2n+1)!} = c_0 \cos x + c_1 \sin x.$$

24. The auxiliary equation is $2m^2 + m = m(2m + 1) = 0$, so $y = c_1 + c_2 e^{-x/2}$. Substituting $y = \sum_{n=0}^{\infty} c_n x^n$ into the differential equation leads to

$$2y'' + y' = 2 \underbrace{\sum_{\substack{n=2}}^{\infty} n(n-1)c_n x^{n-2}}_{k=n-2} + \underbrace{\sum_{n=1}^{\infty} n c_n x^{n-1}}_{k=n-1}$$

$$= 2\sum_{k=0}^{\infty} (k+2)(k+1)c_{k+2}x^k + \sum_{k=0}^{\infty} (k+1)c_{k+1}x^k$$

$$= \sum_{k=0}^{\infty} [2(k+2)(k+1)c_{k+2} + (k+1)c_{k+1}]x^k = 0.$$

Thus
$$2(k+2)(k+1)c_{k+2} + (k+1)c_{k+1} = 0$$

and
$$c_{k+2} = -\frac{1}{2(k+2)} c_{k+1}, \quad k = 0, 1, 2, \ldots.$$

Iterating we find

$$c_2 = -\frac{1}{2}\frac{1}{2} c_1$$

$$c_3 = -\frac{1}{2}\frac{1}{3} c_2 = \frac{1}{2^2}\frac{1}{3 \cdot 2} c_1$$

$$c_4 = -\frac{1}{2}\frac{1}{4} c_3 = \frac{1}{2^3}\frac{1}{4!} c_1$$

and so on. Therefore

$$y = c_0 + c_1 x - \frac{1}{2}\frac{1}{2} c_1 x^2 + \frac{1}{2^2 3!} c_1 x^3 - \frac{1}{2^3 4!} c_1 x^4 + \cdots$$

$$\boxed{\begin{array}{l} c_0 = C_0 - 2c_1 \\ c_1 = -\frac{1}{2}C_1 \end{array}}$$

$$= C_0 + \left[C_1 - \frac{1}{2}C_1 x + \frac{1}{2}\frac{1}{2}\frac{1}{2}C_1 x^2 - \frac{1}{2^2 2 \cdot 3!}\frac{1}{2}C_1 x^3 + \cdots \right]$$

$$= C_0 + C_1 \left[1 - \frac{x}{2} + \frac{1}{2}\left(\frac{x}{2}\right)^2 - \frac{1}{3!}\left(\frac{x}{3}\right)^3 + \cdots \right]$$

$$= C_0 + C_1 \sum_{n=0}^{\infty} \frac{(-1)^n}{n!} \left(\frac{x}{2}\right)^n = C_0 + C_1 \sum_{n=0}^{\infty} \frac{1}{n!}\left(-\frac{x}{n}\right)^n = C_0 + C_1 e^{-x/2}.$$

3. Substituting $y = \sum_{n=0}^{\infty} c_n x^n$ into the differential equation we have

$$y'' - 2xy' + y = \underbrace{\sum_{n=2}^{\infty} n(n-1)c_n x^{n-2}}_{k=n-2} - 2\underbrace{\sum_{n=1}^{\infty} nc_n x^n}_{k=n} + \underbrace{\sum_{n=0}^{\infty} c_n x^n}_{k=n}$$

$$= \sum_{k=0}^{\infty} (k+2)(k+1)c_{k+2}x^k - 2\sum_{k=1}^{\infty} kc_k x^k + \sum_{k=0}^{\infty} c_k x^k$$

$$= 2c_2 + c_0 + \sum_{k=1}^{\infty} [(k+2)(k+1)c_{k+2} - (2k-1)c_k]x^k = 0.$$

Thus
$$2c_2 + c_0 = 0$$

$$(k+2)(k+1)c_{k+2} - (2k-1)c_k = 0$$

and
$$c_2 = -\frac{1}{2}c_0$$

$$c_{k+2} = \frac{2k-1}{(k+2)(k+1)}c_k, \quad k = 1, 2, 3, \ldots .$$

Choosing $c_0 = 1$ and $c_1 = 0$ we find

$$c_2 = -\frac{1}{2}$$

$$c_3 = c_5 = c_7 = \cdots = 0$$

$$c_4 = -\frac{1}{8}$$

$$c_6 = -\frac{7}{336}$$

and so on. For $c_0 = 0$ and $c_1 = 1$ we obtain

$$c_2 = c_4 = c_6 = \cdots = 0$$

$$c_3 = \frac{1}{6}$$

$$c_5 = \frac{1}{24}$$

$$c_7 = \frac{1}{112}$$

64

and so on. Thus, two solutions are

$$y_1 = 1 - \frac{1}{2}x^2 - \frac{1}{8}x^4 - \frac{7}{336}x^6 - \cdots \quad \text{and} \quad y_2 = x + \frac{1}{6}x^3 + \frac{1}{24}x^5 + \frac{1}{112}x^7 + \cdots .$$

6. Substituting $y = \sum_{n=0}^{\infty} c_n x^n$ into the differential equation we have

$$y'' + 2xy' + 2y = \underbrace{\sum_{n=2}^{\infty} n(n-1)c_n x^{n-2}}_{k=n-2} + 2\underbrace{\sum_{n=1}^{\infty} nc_n x^n}_{k=n} + 2\underbrace{\sum_{n=0}^{\infty} c_n x^n}_{k=n}$$

$$= \sum_{k=0}^{\infty}(k+2)(k+1)c_{k+2}x^k + 2\sum_{k=1}^{\infty} kc_k x^k + 2\sum_{k=0}^{\infty} c_k x^k$$

$$= 2c_2 + 2c_0 + \sum_{k=1}^{\infty}[(k+2)(k+1)c_{k+2} + 2(k+1)c_k]x^k = 0.$$

Thus

$$2c_2 + 2c_0 = 0$$

$$(k+2)(k+1)c_{k+2} + 2(k+1)c_k = 0$$

and

$$c_2 = -c_0$$

$$c_{k+2} = -\frac{2}{k+2}c_k, \quad k = 1, 2, 3, \ldots .$$

Choosing $c_0 = 1$ and $c_1 = 0$ we find

$$c_2 = -1$$

$$c_3 = c_5 = c_7 = \cdots = 0$$

$$c_4 = \frac{1}{2}$$

$$c_6 = -\frac{1}{6}$$

and so on. For $c_0 = 0$ and $c_1 = 1$ we obtain

$$c_2 = c_4 = c_6 = \cdots = 0$$

$$c_3 = -\frac{2}{3}$$

$$c_5 = \frac{4}{15}$$

$$c_7 = -\frac{8}{105}$$

65

and so on. Thus, two solutions are

$$y_1 = 1 - x^2 + \frac{1}{2}x^4 - \frac{1}{6}x^6 + \cdots \quad \text{and} \quad y_2 = x - \frac{2}{3}x^3 + \frac{4}{15}x^5 - \frac{8}{105}x^7 + \cdots .$$

9. Substituting $y = \sum_{n=0}^{\infty} c_n x^n$ into the differential equation we have

$$\left(x^2 - 1\right)y'' + 4xy' + 2y = \underbrace{\sum_{n=2}^{\infty} n(n-1)c_n x^n}_{k=n} - \underbrace{\sum_{n=2}^{\infty} n(n-1)c_n x^{n-2}}_{k=n-2} + 4\underbrace{\sum_{n=1}^{\infty} nc_n x^n}_{k=n} + 2\underbrace{\sum_{n=0}^{\infty} c_n x^n}_{k=n}$$

$$= \sum_{k=2}^{\infty} k(k-1)c_k x^k - \sum_{k=0}^{\infty} (k+2)(k+1)c_{k+2}x^k + 4\sum_{k=1}^{\infty} kc_k x^k + 2\sum_{k=0}^{\infty} c_k x^k$$

$$= -2c_2 + 2c_0 + (-6c_3 + 6c_1)x + \sum_{k=2}^{\infty} \left[\left(k^2 - k + 4k + 2\right)c_k - (k+2)(k+1)c_{k+2}\right]x^k = 0.$$

Thus
$$-2c_2 + 2c_0 = 0$$

$$-6c_3 + 6c_1 = 0$$

$$\left(k^2 + 3k + 2\right)c_k - (k+2)(k+1)c_{k+2} = 0$$

and
$$c_2 = c_0$$

$$c_3 = c_1$$

$$c_{k+2} = c_k, \quad k = 2, 3, 4, \ldots .$$

Choosing $c_0 = 1$ and $c_1 = 0$ we find

$$c_2 = 1$$

$$c_3 = c_5 = c_7 = \cdots = 0$$

$$c_4 = c_6 = c_8 = \cdots = 1.$$

For $c_0 = 0$ and $c_1 = 1$ we obtain

$$c_2 = c_4 = c_6 = \cdots = 0$$

$$c_3 = c_5 = c_7 = \cdots = 1.$$

Thus, two solutions are

$$y_1 = 1 + x^2 + x^4 + \cdots \quad \text{and} \quad y_2 = x + x^3 + x^5 + \cdots .$$

12. Substituting $y = \sum_{n=0}^{\infty} c_n x^n$ into the differential equation we have

$$\left(x^2 - 1\right) y'' + xy' - y = \sum_{n=2}^{\infty} n(n-1)c_n x^n - \sum_{n=2}^{\infty} n(n-1)c_n x^{n-2} + \sum_{n=1}^{\infty} nc_n x^n - \sum_{n=0}^{\infty} c_n x^n$$

$$= \sum_{k=2}^{\infty} k(k-1)c_k x^k - \sum_{k=0}^{\infty} (k+2)(k+1)c_{k+2} x^k + \sum_{k=1}^{\infty} kc_k x^k - \sum_{k=0}^{\infty} c_k x^k$$

$$= (-c_2 - c_0) - 6c_3 x + \sum_{k=2}^{\infty} \left[-(k+2)(k+1)c_{k+2} + \left(k^2 - 1\right) c_k\right] x^k = 0.$$

Thus
$$-2c_2 - c_0 = 0$$

$$-6c_3 = 0$$

$$-(k+2)(k+1)c_{k+2} + (k-1)(k+1)c_k = 0$$

and
$$c_2 = -\frac{1}{2}c_0$$

$$c_3 = 0$$

$$c_{k+2} = \frac{k-1}{k+2} c_k, \quad k = 2, 3, 4, \ldots.$$

Choosing $c_0 = 1$ and $c_1 = 0$ we find

$$c_2 = -\frac{1}{2}$$

$$c_3 = c_5 = c_7 = \cdots = 0$$

$$c_4 = -\frac{1}{8}$$

and so on. For $c_0 = 0$ and $c_1 = 1$ we obtain

$$c_2 = c_4 = c_6 = \cdots = 0$$

$$c_3 = c_5 = c_7 = \cdots = 0.$$

Thus, two solutions are

$$y_1 = 1 - \frac{1}{2}x^2 - \frac{1}{8}x^4 - \cdots \quad \text{and} \quad y_2 = x.$$

67

15. Substituting $y = \sum_{n=0}^{\infty} c_n x^n$ into the differential equation we have

$$(x-1)y'' - xy' + y = \underbrace{\sum_{n=2}^{\infty} n(n-1)c_n x^{n-1}}_{k=n-1} - \underbrace{\sum_{n=2}^{\infty} n(n-1)c_n x^{n-2}}_{k=n-2} - \underbrace{\sum_{n=1}^{\infty} nc_n x^n}_{k=n} + \underbrace{\sum_{n=0}^{\infty} c_n x^n}_{k=n}$$

$$= \sum_{k=1}^{\infty} (k+1)kc_{k+1}x^k - \sum_{k=0}^{\infty} (k+2)(k+1)c_{k+2}x^k - \sum_{k=1}^{\infty} kc_k x^k + \sum_{k=0}^{\infty} c_k x^k$$

$$= -2c_2 + c_0 + \sum_{k=1}^{\infty} [-(k+2)(k+1)c_{k+2} + (k+1)kc_{k+1} - (k-1)c_k]x^k = 0.$$

Thus
$$-2c_2 + c_0 = 0$$

$$-(k+2)(k+1)c_{k+2} + (k-1)kc_{k+1} - (k-1)c_k = 0$$

and
$$c_2 = \frac{1}{2}c_0$$

$$c_{k+2} = \frac{kc_{k+1}}{k+2} - \frac{(k-1)c_k}{(k+2)(k+1)}, \qquad k = 1, 2, 3, \ldots.$$

Choosing $c_0 = 1$ and $c_1 = 0$ we find

$$c_2 = \frac{1}{2}, \qquad c_3 = \frac{1}{6}, \qquad c_4 = 0$$

and so on. For $c_0 = 0$ and $c_1 = 1$ we obtain $c_2 = c_3 = c_4 = \cdots = 0$. Thus,

$$y = C_1\left(1 + \frac{1}{2}x^2 + \frac{1}{6}x^3 + \cdots\right) + C_2 x$$

and
$$y' = C_1\left(x + \frac{1}{2}x^2 + \cdots\right) + C_2.$$

The initial conditions imply $C_1 = -2$ and $C_2 = 6$, so

$$y = -2\left(1 + \frac{1}{2}x^2 + \frac{1}{6}x^3 + \cdots\right) + 6x = 8x - 2e^x.$$

18. Substituting $y = \sum_{n=0}^{\infty} c_n x^n$ into the differential equation we have

$$(x^2 + 1)y'' + 2xy' = \sum_{n=2}^{\infty} \underbrace{n(n-1)c_n x^n}_{k=n} + \sum_{n=2}^{\infty} \underbrace{n(n-1)c_n x^{n-2}}_{k=n-2} + \sum_{n=1}^{\infty} \underbrace{2nc_n x^n}_{k=n}$$

$$= \sum_{k=2}^{\infty} k(k-1)c_k x^k + \sum_{k=0}^{\infty} (k+2)(k+1)c_{k+2} x^k + \sum_{k=1}^{\infty} 2kc_k x^k$$

$$= 2c_2 + (6c_3 + 2c_1)x + \sum_{k=2}^{\infty} [k(k+1)c_k + (k+2)(k+1)c_{k+2}]x^k = 0.$$

Thus
$$2c_2 = 0$$

$$6c_3 + 2c_1 = 0$$

$$k(k+1)c_k + (k+2)(k+1)c_{k+2} = 0$$

and
$$c_2 = 0$$

$$c_3 = -\frac{1}{3}c_1$$

$$c_{k+2} = -\frac{k}{k+2}c_k, \quad k = 2, 3, 4, \ldots.$$

Choosing $c_0 = 1$ and $c_1 = 0$ we find $c_3 = c_4 = c_5 = \cdots = 0$. For $c_0 = 0$ and $c_1 = 1$ we obtain

$$c_3 = -\frac{1}{3}$$

$$c_4 = c_6 = c_8 = \cdots = 0$$

$$c_5 = -\frac{1}{5}$$

$$c_7 = \frac{1}{7}$$

and so on. Thus

$$y = c_0 + c_1 \left(x - \frac{1}{3}x^3 + \frac{1}{5}x^5 - \frac{1}{7}x^7 + \cdots \right)$$

and

$$y' = c_1 \left(1 - x^2 + x^4 - x^6 + \cdots \right).$$

The initial conditions imply $c_0 = 0$ and $c_1 = 1$, so

$$y = x - \frac{1}{3}x^3 + \frac{1}{5}x^5 - \frac{1}{7}x^7 + \cdots.$$

69

21. Substituting $y = \sum_{n=0}^{\infty} c_n x^n$ into the differential equation we have

$$y'' + e^{-x} y = \sum_{n=2}^{\infty} n(n-1) c_n x^{n-2}$$

$$+ \left(1 - x + \frac{1}{2} x^2 - \frac{1}{6} x^3 + \frac{1}{24} x^4 - \cdots \right) \left(c_0 + c_1 x + c_2 x^2 + c_3 x^3 + \cdots \right)$$

$$= \left[2c_2 + 6c_3 x + 12 c_4 x^2 + 20 c_5 x^3 + \cdots \right] + \left[c_0 + (c_1 - c_0) x + \left(c_2 - c_1 + \frac{1}{2} c_0 \right) x^2 + \cdots \right]$$

$$= (2c_2 + c_0) + (6c_3 + c_1 - c_0) x + \left(12 c_4 + c_2 - c_1 + \frac{1}{2} c_0 \right) x^2 + \cdots = 0.$$

Thus
$$2c_2 + c_0 = 0$$

$$6c_3 + c_1 - c_0 = 0$$

$$12 c_4 + c_2 - c_1 + \frac{1}{2} c_0 = 0$$

and
$$c_2 = -\frac{1}{2} c_0$$

$$c_3 = -\frac{1}{6} c_1 + \frac{1}{6} c_0$$

$$c_4 = -\frac{1}{12} c_2 + \frac{1}{12} c_1 - \frac{1}{24} c_0.$$

Choosing $c_0 = 1$ and $c_1 = 0$ we find

$$c_2 = -\frac{1}{2}, \qquad c_3 = \frac{1}{6}, \qquad c_4 = 0$$

and so on. For $c_0 = 0$ and $c_1 = 1$ we obtain

$$c_2 = 0, \qquad c_3 = -\frac{1}{6}, \qquad c_4 = \frac{1}{12}.$$

Thus, two solutions are

$$y_1 = 1 - \frac{1}{2} x^2 + \frac{1}{6} x^3 + \cdots \qquad \text{and} \qquad y_2 = x - \frac{1}{6} x^3 + \frac{1}{12} x^4 + \cdots.$$

24. Substituting $y = \sum_{n=0}^{\infty} c_n x^n$ into the differential equation leads to

$$y'' - 4xy' - 4y = \underbrace{\sum_{n=2}^{\infty} n(n-1) c_n x^{n-2}}_{k=n-2} - \underbrace{\sum_{n=1}^{\infty} 4n c_n x^n}_{k=n} - \underbrace{\sum_{n=0}^{\infty} 4 c_n x^n}_{k=n}$$

$$= \sum_{k=0}^{\infty} (k+2)(k+1) c_{k+2} x^k - \sum_{k=1}^{\infty} 4k c_k x^k - \sum_{k=0}^{\infty} 4 c_k x^k$$

$$= 2c_2 - 4c_0 + \sum_{k=1}^{\infty} [(k+2)(k+1)c_{k+2} - 4(k+1)c_k]x^k$$

$$= e^x = 1 + \sum_{k-1}^{\infty} \frac{1}{k!}x^k.$$

Thus
$$2c_2 - 4c_0 = 1$$

$$(k+2)(k+1)c_{k+2} - 4(k+1)c_k = \frac{1}{k!}$$

and
$$c_2 = \frac{1}{2} + 2c_0$$

$$c_{k+2} = \frac{1}{(k+2)!} + \frac{4}{k+2}c_k, \qquad k = 1, 2, 3, \ldots .$$

Let c_0 and c_1 be arbitrary and iterate to find

$$c_2 = \frac{1}{2} + 2c_0$$

$$c_3 = \frac{1}{3!} + \frac{4}{3}c_1 = \frac{1}{3!} + \frac{4}{3}c_1$$

$$c_4 = \frac{1}{4!} + \frac{4}{4}c_2 = \frac{1}{4!} + \frac{1}{2} + 2c_0 = \frac{13}{4!} + 2c_0$$

$$c_5 = \frac{1}{5!} + \frac{4}{5}c_3 = \frac{1}{5!} + \frac{4}{5 \cdot 3!} + \frac{16}{15}c_1 = \frac{17}{5!} + \frac{16}{15}c_1$$

$$c_6 = \frac{1}{6!} + \frac{4}{6}c_4 = \frac{1}{6!} + \frac{4 \cdot 13}{6 \cdot 4!} + \frac{8}{6}c_0 = \frac{261}{6!} + \frac{4}{3}c_0$$

$$c_7 - \frac{1}{7!} + \frac{4}{7}c_5 = \frac{1}{7!} + \frac{4 \cdot 17}{7 \cdot 5!} + \frac{64}{105}c_1 = \frac{409}{7!} + \frac{64}{105}c_1$$

and so on. The solution is

$$y = c_0 + c_1 x + \left(\frac{1}{2} + 2c_0\right)x^2 + \left(\frac{1}{3!} + \frac{4}{3}c_1\right)x^3 - \left(\frac{13}{4!} + 2c_0\right)x^4 + \left(\frac{17}{5!} + \frac{16}{15}c_1\right)x^5$$

$$+ \left(\frac{261}{6!} + \frac{4}{3}c_0\right)x^6 + \left(\frac{409}{7!} + \frac{64}{105}c_1\right)x^7 + \cdots$$

$$= c_0\left[1 + 2x^2 + 2x^4 + \frac{4}{3}x^6 + \cdots\right] + c_1\left[x + \frac{4}{3}x^3 + \frac{16}{15}x^5 + \frac{64}{105}x^7 + \cdots\right]$$

$$+ \frac{1}{2}x^2 + \frac{1}{3!}x^3 + \frac{13}{4!}x^4 + \frac{17}{5!}x^5 + \frac{261}{6!}x^6 + \frac{409}{7!}x^7 + \cdots .$$

—————— **Exercises 6.3** ——————————————

3. Irregular singular point: $x = 3$; regular singular point: $x = -3$

6. Irregular singular point: $x = 5$; regular singular point: $x = 0$

9. Irregular singular point: $x = 0$; regular singular points: $x = 2, \pm 5$

12. Substituting $y = \sum_{n=0}^{\infty} c_n x^{n+r}$ into the differential equation and collecting terms, we obtain

$$2xy'' + 5y' + xy = \left(2r^2 + 3r\right) c_0 x^{r-1} + \left(2r^2 + 7r + 5\right) c_1 x^r$$

$$+ \sum_{k=2}^{\infty} [2(k+r)(k+r-1)c_k + 5(k+r)c_k + c_{k-2}]x^{k+r-1}$$

$$= 0,$$

which implies

$$2r^2 + 3r = r(2r + 3) = 0,$$

$$\left(2r^2 + 7r + 5\right) c_1 = 0,$$

and $\qquad\qquad (k+r)(2k + 2r + 3)c_k + c_{k-2} = 0.$

The indicial roots are $r = -3/2$ and $r = 0$, so $c_1 = 0$. For $r = -3/2$ the recurrence relation is

$$c_k = -\frac{c_{k-2}}{(2k-3)k}, \quad k = 2, 3, 4, \ldots,$$

and $\qquad\qquad c_2 = -\frac{1}{2}c_0, \qquad c_3 = 0, \qquad c_4 = \frac{1}{40}c_0.$

For $r = 0$ the recurrence relation is

$$c_k = -\frac{c_{k-2}}{k(2k+3)}, \quad k = 2, 3, 4, \ldots,$$

and $\qquad\qquad c_2 = -\frac{1}{14}c_0, \qquad c_3 = 0, \qquad c_4 = \frac{1}{616}c_0.$

The general solution on $(0, \infty)$ is

$$y = C_1 x^{-3/2} \left(1 - \frac{1}{2}x^2 + \frac{1}{40}x^4 + \cdots\right) + C_2 \left(1 - \frac{1}{14}x^2 + \frac{1}{616}x^4 + \cdots\right).$$

15. Substituting $y = \sum_{n=0}^{\infty} c_n x^{n+r}$ into the differential equation and collecting terms, we obtain

$$3xy'' + (2 - x)y' - y = \left(3r^2 - r\right) c_0 x^{r-1}$$

$$+ \sum_{k=1}^{\infty} [3(k+r-1)(k+r)c_k + 2(k+r)c_k - (k+r)c_{k-1}]x^{k+r-1}$$

$$= 0,$$

which implies $$3r^2 - r = r(3r-1) = 0$$

and $$(k+r)(3k+3r-1)c_k - (k+r)c_{k-1} = 0.$$

The indicial roots are $r=0$ and $r=1/3$. For $r=0$ the recurrence relation is

$$c_k = \frac{c_{k-1}}{(3k-1)}, \quad k=1,2,3,\ldots,$$

and $$c_1 = \frac{1}{2}c_0, \quad c_2 = \frac{1}{10}c_0, \quad c_3 = \frac{1}{80}c_0.$$

For $r=1/3$ the recurrence relation is

$$c_k = \frac{c_{k-1}}{3k}, \quad k=1,2,3,\ldots,$$

and $$c_1 = \frac{1}{3}c_0, \quad c_2 = \frac{1}{18}c_0, \quad c_3 = \frac{1}{162}c_0.$$

The general solution on $(0,\infty)$ is

$$y = C_1\left(1 + \frac{1}{2}x + \frac{1}{10}x^2 + \frac{1}{80}x^3 + \cdots\right) + C_2 x^{1/3}\left(1 + \frac{1}{3}x + \frac{1}{18}x^2 + \frac{1}{162}x^3 + \cdots\right).$$

18. Substituting $y = \sum_{n=0}^\infty c_n x^{n+r}$ into the differential equation and collecting terms, we obtain

$$x^2 y'' + xy' + \left(x^2 - \frac{4}{9}\right)y = \left(r^2 - \frac{4}{9}\right)c_0 x^r + \left(r^2 + 2r + \frac{5}{9}\right)c_1 x^{r+1}$$

$$+ \sum_{k=2}^\infty \left[(k+r)(k+r-1)c_k + (k+r)c_k - \frac{4}{9}c_k + c_{k-2}\right]x^{k+r}$$

$$= 0,$$

which implies

$$r^2 - \frac{4}{9} = \left(r + \frac{2}{3}\right)\left(r - \frac{2}{3}\right) = 0,$$

$$\left(r^2 + 2r + \frac{5}{9}\right)c_1 = 0,$$

and $$\left[(k+r)^2 - \frac{4}{9}\right]c_k + c_{k-2} = 0.$$

The indicial roots are $r=-2/3$ and $r=2/3$, so $c_1=0$. For $r=-2/3$ the recurrence relation is

$$c_k = -\frac{9c_{k-2}}{3k(3k-4)}, \quad k=2,3,4,\ldots,$$

and $$c_2 = -\frac{3}{4}c_0, \quad c_3 = 0, \quad c_4 = \frac{9}{128}c_0.$$

For $r = 2/3$ the recurrence relation is

$$c_k = -\frac{9c_{k-2}}{3k(3k+4)}, \quad k = 2, 3, 4, \ldots,$$

and

$$c_2 = -\frac{3}{20}c_0, \quad c_3 = 0, \quad c_4 = \frac{9}{1,280}c_0.$$

The general solution on $(0, \infty)$ is

$$y = C_1 x^{-2/3}\left(1 - \frac{3}{4}x^2 + \frac{9}{128}x^4 + \cdots\right) + C_2 x^{2/3}\left(1 - \frac{3}{20}x^2 + \frac{9}{1,280}x^4 + \cdots\right).$$

21. Substituting $y = \sum_{n=0}^{\infty} c_n x^{n+r}$ into the differential equation and collecting terms, we obtain

$$2x^2 y'' - x(x-1)y' - y = \left(2r^2 - r - 1\right)c_0 x^r$$

$$+ \sum_{k=1}^{\infty}[2(k+r)(k+r-1)c_k + (k+r)c_k - c_k - (k+r-1)c_{k-1}]x^{k+r}$$

$$= 0,$$

which implies

$$2r^2 - r - 1 = (2r+1)(r-1) = 0$$

and

$$[(k+r)(2k+2r-1)-1]c_k - (k+r-1)2c_{k-1} = 0.$$

The indicial roots are $r = -1/2$ and $r = 1$. For $r = -1/2$ the recurrence relation is

$$c_k = \frac{c_{k-1}}{2k}, \quad k = 1, 2, 3, \ldots,$$

and

$$c_1 = \frac{1}{2}c_0, \quad c_2 = \frac{1}{8}c_0, \quad c_3 = \frac{1}{48}c_0.$$

For $r = 1$ the recurrence relation is

$$c_k = \frac{c_{k-1}}{2k+3}, \quad k = 1, 2, 3, \ldots,$$

and

$$c_1 = \frac{1}{5}c_0, \quad c_2 = \frac{1}{35}c_0, \quad c_3 = \frac{1}{315}c_0.$$

The general solution on $(0, \infty)$ is

$$y = C_1 x^{-1/2}\left(1 + \frac{1}{2}x + \frac{1}{8}x^2 + \frac{1}{48}x^3 + \cdots\right) + C_2 x\left(1 + \frac{1}{5}x + \frac{1}{35}x^2 + \frac{1}{315}x^3 + \cdots\right).$$

24. Substituting $y = \sum_{n=0}^{\infty} c_n x^{n+r}$ into the differential equation and collecting terms, we obtain

$$x^2 y'' + xy' + \left(x^2 - \frac{1}{4}\right)y = \left(r^2 - \frac{1}{4}\right)c_0 x^r + \left(r^2 + 2r + \frac{3}{4}\right)c_1 x^{r+1}$$

$$+ \sum_{k=2}^{\infty}\left[(k+r)(k+r-1)c_k + (k+r)c_k - \frac{1}{4}c_k + c_{k-2}\right]x^{k+r}$$

$$= 0,$$

74

which implies

$$r^2 - \frac{1}{4} = \left(r - \frac{1}{2} \right) \left(r + \frac{1}{2} \right) = 0,$$

$$\left(r^2 + 2r + \frac{3}{4} \right) c_1 = 0,$$

and

$$\left[(k + r)^2 - \frac{1}{4} \right] c_k + c_{k-2} = 0.$$

The indicial roots are $r_1 = 1/2$ and $r_2 = -1/2$, so $c_1 = 0$. For $r_1 = 1/2$ the recurrence relation is

$$c_k = -\frac{c_{k-2}}{k(k+1)}, \quad k = 2, 3, 4, \dots,$$

and

$$c_2 = -\frac{1}{3!} c_0$$

$$c_3 = c_5 = c_7 = \cdots = 0$$

$$c_4 = \frac{1}{5!} c_0$$

$$c_{2n} = \frac{(-1)^n}{(2n+1)!} c_0.$$

For $r_2 = -1/2$ the recurrence relation is

$$c_k = -\frac{c_{k-2}}{k(k-1)}, \quad k = 2, 3, 4, \dots,$$

and

$$c_2 = -\frac{1}{2!} c_0$$

$$c_3 = c_5 = c_7 = \cdots = 0$$

$$c_4 = \frac{1}{4!} c_0$$

$$c_{2n} = \frac{(-1)^n}{(2n)!} c_0.$$

The general solution on $(0, \infty)$ is

$$y = C_1 x^{1/2} \sum_{n=0}^{\infty} \frac{(-1)^n}{(2n+1)!} x^{2n} + C_2 x^{-1/2} \sum_{n=0}^{\infty} \frac{(-1)^n}{(2n)!} x^{2n}$$

$$= C_1 x^{-1/2} \sum_{n=0}^{\infty} \frac{(-1)^n}{(2n+1)!} x^{2n+1} + C_2 x^{-1/2} \sum_{n=0}^{\infty} \frac{(-1)^n}{(2n)!} x^{2n}$$

$$= x^{-1/2} [C_1 \sin x + C_2 \cos x].$$

75

Exercises 6.3

27. Substituting $y = \sum_{n=0}^{\infty} c_n x^{n+r}$ into the differential equation and collecting terms, we obtain

$$xy'' + (1-x)y' - y = r^2 c_0 x^{r-1} + \sum_{k=0}^{\infty} [(k+r)(k+r-1)c_k + (k+r)c_k - (k+r)c_{k-1}]x^{k+r-1} = 0,$$

which implies $r^2 = 0$ and

$$(k+r)^2 c_k - (k+r)c_{k-1} = 0.$$

The indicial roots are $r_1 = r_2 = 0$ and the recurrence relation is

$$c_k = \frac{c_{k-1}}{k}, \quad k = 1, 2, 3, \ldots .$$

One solution is

$$y_1 = c_0 \left(1 + x + \frac{1}{2}x^2 + \frac{1}{3!}x^3 + \cdots \right) = c_0 e^x.$$

A second solution is

$$y_2 = y_1 \int \frac{e^{-\int (1/x - 1)dx}}{e^{2x}}\, dx = e^x \int \frac{e^x/x}{e^{2x}}\, dx = e^x \int \frac{1}{x}e^{-x}dx$$

$$= e^x \int \frac{1}{x}\left(1 - x + \frac{1}{2}x^2 - \frac{1}{3!}x^3 + \cdots \right)dx = e^x \int \left(\frac{1}{x} - 1 + \frac{1}{2}x - \frac{1}{3!}x^2 + \cdots \right)dx$$

$$= e^x \left[\ln x - x + \frac{1}{2 \cdot 2}x^2 - \frac{1}{3 \cdot 3!}x^3 + \cdots \right] = e^x \ln x - e^x \sum_{n=1}^{\infty} \frac{(-1)^{n+1}}{n \cdot n!}x^n.$$

The general solution on $(0, \infty)$ is

$$y = C_1 e^x + C_2 e^x \left(\ln x - \sum_{n=1}^{\infty} \frac{(-1)^{n+1}}{n \cdot n!}x^n \right).$$

30. Substituting $y = \sum_{n=0}^{\infty} c_n x^{n+r}$ into the differential equation and collecting terms, we obtain

$$xy'' - xy' + y = \left(r^2 - r \right) c_0 x^{r-1} + \sum_{k=0}^{\infty} [(k+r+1)(k+r)c_{k+1} - (k+r)c_k + c_k]x^{k+r} = 0$$

which implies

$$r^2 - r = r(r-1) = 0$$

and

$$(k+r+1)(k+r)c_{k+1} - (k+r-1)c_k = 0.$$

The indicial roots are $r_1 = 1$ and $r_2 = 0$. For $r_1 = 1$ the recurrence relation is

$$c_{k+1} = \frac{kc_k}{(k+2)(k+1)}, \quad k = 0, 1, 2, \ldots,$$

76

and one solution is $y_1 = c_0 x$. A second solution is

$$y_2 = x \int \frac{e^{-\int -dx}}{x^2} \, dx = x \int \frac{e^x}{x^2} \, dx = x \int \frac{1}{x^2} \left(1 + x + \frac{1}{2} x^2 + \frac{1}{3!} x^3 + \cdots \right) dx$$

$$= x \int \left(\frac{1}{x^2} + \frac{1}{x} + \frac{1}{2} + \frac{1}{3!} x + \frac{1}{4!} x^2 + \cdots \right) dx = x \left[-\frac{1}{x} + \ln x + \frac{1}{2} x + \frac{1}{12} x^2 + \frac{1}{72} x^3 + \cdots \right]$$

$$= x \ln x - 1 + \frac{1}{2} x^2 + \frac{1}{12} x^3 + \frac{1}{72} x^4 + \cdots .$$

The general solution on $(0, \infty)$ is

$$y = C_1 x + C_2 y_2(x).$$

Exercises 6.4

3. Since $\nu^2 = 25/4$ the general solution is $y = c_1 J_{5/2}(x) + c_2 J_{-5/2}(x)$.

6. Since $\nu^2 = 4$ the general solution is $y = c_1 J_2(x) + c_2 Y_2(x)$.

9. If $y = x^{-1/2} v(x)$ then

$$y' = x^{-1/2} v'(x) - \frac{1}{2} x^{-3/2} v(x),$$

$$y'' = x^{-1/2} v''(x) - x^{-3/2} v'(x) + \frac{3}{4} x^{-5/2} v(x),$$

and

$$x^2 y'' + 2xy' + \lambda^2 x^2 y = x^{3/2} v'' + x^{1/2} v' + \left(\lambda^2 x^{3/2} - \frac{1}{4} x^{-1/2} \right) v.$$

Multiplying by $x^{1/2}$ we obtain

$$x^2 v'' + x v' + \left(\lambda^2 x^2 - \frac{1}{4} \right) v = 0,$$

whose solution is $v = c_1 J_{1/2}(\lambda x) + c_2 J_{-1/2}(\lambda x)$. Then $y = c_1 x^{-1/2} J_{1/2}(\lambda x) + c_2 x^{-1/2} J_{-1/2}(\lambda x)$.

12. From $y = \sqrt{x} \, J_\nu(\lambda x)$ we find

$$y' = \lambda \sqrt{x} \, J_\nu'(\lambda x) + \frac{1}{2} x^{-1/2} J_\nu(\lambda x)$$

and

$$y'' = \lambda^2 \sqrt{x} \, J_\nu''(\lambda x) + \lambda x^{-1/2} J_\nu'(\lambda x) - \frac{1}{4} x^{-3/2} J_\nu(\lambda x).$$

Substituting into the differential equation, we have

$$x^2 y'' + \left(\lambda^2 x^2 - \nu^2 + \frac{1}{4}\right) y = \sqrt{x}\, \left[\lambda^2 x^2 J_\nu''(\lambda x) + \lambda x J_\nu'(\lambda x) + \left(\lambda^2 x^2 - \nu^2\right) J_\nu(\lambda x)\right]$$

$$= \sqrt{x} \cdot 0 \qquad \text{(since } J_n \text{ is a solution of Bessel's equation)}$$

$$= 0.$$

Therefore, $\sqrt{x}\, J_\nu(\lambda x)$ is a solution of the original equation.

15. From Problem 10 with $n = -1$ we find $y = x^{-1} J_{-1}(x)$. From Problem 11 with $n = 1$ we find $y = x^{-1} J_1(x) = -x^{-1} J_{-1}(x)$.

18. From Problem 10 with $n = 3$ we find $y = x^3 J_3(x)$. From Problem 11 with $n = -3$ we find $y = x^3 J_{-3}(x) = -x^3 J_3(x)$.

21. The recurrence relation follows from

$$x J_{\nu+1}(x) + x J_{\nu-1}(x) = \sum_{n=0}^{\infty} \frac{(-1)^{n-1} 2n}{n!\,\Gamma(1+\nu+n)} \left(\frac{x}{2}\right)^{2n+\nu} + \sum_{n=0}^{\infty} \frac{(-1)^n 2(\nu+n)}{n!\,\Gamma(1+\nu+n)} \left(\frac{x}{2}\right)^{2n+\nu}$$

$$= \sum_{n=0}^{\infty} \frac{(-1)^n 2\nu}{n!\,\Gamma(1+\nu+n)} \left(\frac{x}{2}\right)^{2n+\nu} = 2\nu J_\nu(x).$$

24. From (14) in the text we obtain $J_0'(x) = -J_1(x)$ and from (15) in the text we obtain $J_0'(x) = J_{-1}(x)$. Thus

$$J_0'(x) = J_{-1}(x) = -J_1(x).$$

27. Since

$$\Gamma\left(1 - \frac{1}{2} + n\right) = \frac{(2n-1)!}{(n-1)!\,2^{2n-1}}$$

we obtain

$$J_{-1/2}(x) = \sum_{n=0}^{\infty} \frac{(-1)^n 2^{1/2} x^{-1/2}}{2n(2n-1)!\sqrt{\pi}} x^{2n} = \sqrt{\frac{2}{\pi x}}\, \cos x.$$

30. By Problem 21 we obtain $3 J_{3/2}(x) = x J_{5/2}(x) + x J_{1/2}(x)$ so that

$$J_{5/2}(x) = \sqrt{\frac{2}{\pi x}} \left(\frac{3\sin x}{x^2} - \frac{3\cos x}{x} - \sin x\right).$$

33. By Problem 21 we obtain $-5 J_{-5/2}(x) = x J_{-3/2}(x) + x J_{-7/2}(x)$ so that

$$J_{-7/2}(x) = \sqrt{\frac{2}{\pi x}} \left(\frac{-15\cos x}{x^3} - \frac{15\sin x}{x^2} + \frac{6\cos x}{x} + \sin x\right).$$

36. If $y_1 = J_0(x)$ then using the formula for the second solution of a linear homogeneous second-order differential equation gives

$$y_2 = J_0(x) \int \frac{e^{-\int dx/x}}{(J_0(x))^2} \, dx$$

$$= J_0(x) \int \frac{dx}{x\left(1 - \dfrac{x^2}{4} + \dfrac{x^4}{64} - \dfrac{x^6}{2304} + \cdots\right)^2} \, dx$$

$$= J_0(x) \int \left(\frac{1}{x} + \frac{x}{2} + \frac{5x^3}{32} + \frac{23x^5}{576} + \cdots\right) dx$$

$$= J_0(x) \left(\ln x + \frac{x^2}{4} + \frac{5x^4}{128} + \frac{23x^6}{3456} + \cdots\right)$$

$$= J_0(x) \ln x + \left(1 - \frac{x^2}{4} + \frac{x^4}{64} - \frac{x^6}{2304} + \cdots\right)\left(\frac{x^2}{4} + \frac{5x^4}{128} + \frac{23x^6}{3456} + \cdots\right)$$

$$= J_0(x) \ln x + \frac{x^2}{4} - \frac{3x^4}{128} + \frac{11x^6}{13824} - \cdots \, .$$

39. Letting

$$s = \frac{2}{\alpha}\sqrt{\frac{k}{m}}\, e^{-\alpha t/2},$$

we have

$$\frac{dx}{dt} = \frac{dx}{ds}\frac{ds}{dt} = \frac{dx}{dt}\left[\frac{2}{\alpha}\sqrt{\frac{k}{m}}\left(-\frac{\alpha}{2}\right)e^{-\alpha t/2}\right]$$

$$= \frac{dx}{ds}\left(-\sqrt{\frac{k}{m}}\, e^{-\alpha t/2}\right)$$

and

$$\frac{d^2x}{dt^2} = \frac{d}{dt}\left(\frac{dx}{dt}\right) = \frac{dx}{ds}\left(\frac{\alpha}{2}\sqrt{\frac{k}{m}}\, e^{-\alpha t/2}\right) + \frac{d}{dt}\left(\frac{dx}{ds}\right)\left(-\sqrt{\frac{k}{m}}\, e^{-\alpha t/2}\right)$$

$$= \frac{dx}{ds}\left(\frac{\alpha}{2}\sqrt{\frac{k}{m}}\, e^{-\alpha t/2}\right) + \frac{d^2x}{ds^2}\frac{ds}{dt}\left(-\sqrt{\frac{k}{m}}\, e^{-\alpha t/2}\right)$$

$$= \frac{dx}{ds}\left(\frac{\alpha}{2}\sqrt{\frac{k}{m}}\, e^{-\alpha t/2}\right) + \frac{d^2x}{ds^2}\left(\frac{k}{m}\, e^{-\alpha t}\right).$$

Then

$$m\frac{d^2x}{dt^2} + ke^{-\alpha t}x = ke^{-\alpha t}\frac{d^2x}{ds^2} + \frac{m\alpha}{2}\sqrt{\frac{k}{m}}\, e^{-\alpha t/2}\frac{dx}{dt} + ke^{-\alpha t}x = 0.$$

Multiplying by $2^2/\alpha^2 m$ we have

$$\frac{2^2}{\alpha^2}\frac{k}{m}e^{-\alpha t}\frac{d^2 x}{ds^2} + \frac{2}{\alpha}\sqrt{\frac{k}{m}}e^{-\alpha t/2}\frac{dx}{dt} + \frac{2}{\alpha^2}\frac{k}{m}e^{-\alpha t}x = 0$$

or, since $s = (2/\alpha)\sqrt{k/m}\,e^{-\alpha t/2}$,

$$s^2\frac{d^2 x}{ds^2} + s\frac{dx}{ds} + s^2 x = 0.$$

42. The general solution of Bessel's equation is

$$w(t) = c_1 J_{1/3}(t) + c_2 J_{-1/3}(t), \qquad t > 0.$$

Thus, the general solution of Airy's equation for $x > 0$ is

$$y = x^{1/2}w\left(\frac{2}{3}\alpha x^{3/2}\right) = c_1 x^{1/2}J_{1/3}\left(\frac{2}{3}\alpha x^{3/2}\right) + c_2 x^{1/2}J_{-1/3}\left(\frac{2}{3}\alpha x^{3/2}\right).$$

45. (a) Using the expressions for the two linearly independent power series solutions, $y_1(x)$ and $y_2(x)$, given in the text we obtain

$$P_6(x) = \frac{1}{16}\left(231x^6 - 315x^4 + 105x^2 - 5\right)$$

and

$$P_7(x) = \frac{1}{16}\left(429x^7 - 693x^5 + 315x^3 - 35x\right).$$

(b) $P_6(x)$ satisfies $\left(1 - x^2\right)y'' - 2xy' + 42y = 0$ and $P_7(x)$ satisfies $\left(1 - x^2\right)y'' - 2xy' + 56y = 0$.

48. The polynomials are shown in (19) on page 251 in the text.

51. The recurrence relation can be written

$$P_{k+1}(x) = \frac{2k+1}{k+1}xP_k(x) - \frac{k}{k+1}P_{k-1}(x), \qquad k = 2,\ 3,\ 4,\ \dots\ .$$

$k = 1$: $\quad P_2(x) = \dfrac{3}{2}x^2 - \dfrac{1}{2}$

$k = 2$: $\quad P_3(x) = \dfrac{5}{3}x\left(\dfrac{3}{2}x^2 - \dfrac{1}{2}\right) - \dfrac{2}{3}x = \dfrac{5}{2}x^3 - \dfrac{3}{2}x$

$k = 3$: $\quad P_4(x) = \dfrac{7}{4}x\left(\dfrac{5}{2}x^3 - \dfrac{3}{2}x\right) - \dfrac{3}{4}\left(\dfrac{3}{2}x^2 - \dfrac{1}{2}\right) = \dfrac{35}{8}x^4 - \dfrac{30}{8}x^2 + \dfrac{3}{8}$

$k = 4$: $\quad P_5(x) = \dfrac{9}{5}x\left(\dfrac{35}{8}x^4 - \dfrac{30}{8}x^2 + \dfrac{3}{8}\right) - \dfrac{4}{5}\left(\dfrac{5}{2}x^3 - \dfrac{3}{2}x\right) = \dfrac{63}{8}x^5 - \dfrac{35}{4}x^3 + \dfrac{15}{8}x$

$k = 5$: $\quad P_6(x) = \dfrac{11}{6}x\left(\dfrac{63}{8}x^5 - \dfrac{35}{4}x^3 + \dfrac{15}{8}x\right) - \dfrac{5}{6}\left(\dfrac{35}{8}x^4 - \dfrac{30}{8}x^2 + \dfrac{3}{8}\right) = \dfrac{231}{16}x^6 - \dfrac{315}{16}x^4 + \dfrac{105}{16}x - \dfrac{5}{16}$

Chapter 6 Review Exercises

3. Since

$$P(x) = \frac{1}{x(x-5)^2} \quad \text{and} \quad Q(x) - 0$$

the regular singular point is $x = 0$ and the irregular singular point is $x = 5$.

6. Since

$$P(x) = \frac{1}{x\left(x^2 + 1\right)^3} \quad \text{and} \quad Q(x) = -\frac{8}{\left(x^2 + 1\right)^3}$$

the regular singular point is $x = 0$. The irregular singular points are $x = i$ and $x = -i$.

9. Substituting $y = \sum_{n=0}^{\infty} c_n x^n$ into the differential equation we have

$$y'' - xy' - y = \underbrace{\sum_{n=2}^{\infty} n(n-1)c_n x^{n-2}}_{k=n-2} - \underbrace{\sum_{n=1}^{\infty} nc_n x^n}_{k=n} - \underbrace{\sum_{n=0}^{\infty} c_n x^n}_{k=n}$$

$$= \sum_{k=0}^{\infty} (k+2)(k+1)c_{k+2} x^k - \sum_{k=1}^{\infty} kc_k x^k - \sum_{k=0}^{\infty} c_k x^k$$

$$= 2c_2 - c_0 + \sum_{k=1}^{\infty} [(k+2)(k+1)c_{k+2} - (k+1)c_k]x^k = 0.$$

Thus
$$2c_2 - c_0 = 0$$

$$(k+2)(k+1)c_{k+2} - (k+1)c_k = 0$$

and
$$c_2 = \frac{1}{2}c_0$$

$$c_{k+2} = \frac{1}{k+2}c_k, \quad k = 1, 2, 3, \ldots.$$

Choosing $c_0 = 1$ and $c_1 = 0$ we find

$$c_2 = \frac{1}{2}$$

$$c_3 = c_5 = c_7 = \cdots = 0$$

$$c_4 = \frac{1}{8}$$

$$c_6 = \frac{1}{48}$$

and so on. For $c_0 = 0$ and $c_1 = 1$ we obtain

$$c_2 = c_4 = c_6 = \cdots = 0$$

$$c_3 = \frac{1}{3}$$

$$c_5 = \frac{1}{15}$$

$$c_7 = \frac{1}{105}$$

and so on. Thus, two solutions are

$$y_1 = 1 + \frac{1}{2}x^2 + \frac{1}{8}x^4 + \frac{1}{48}x^6 + \cdots$$

and

$$y_2 = x + \frac{1}{3}x^3 + \frac{1}{15}x^5 + \frac{1}{105}x^7 + \cdots.$$

12. Substituting $y = \sum_{n=0}^{\infty} c_n x^n$ into the differential equation we have

$$(\cos x)y'' + y = \left(1 - \frac{1}{2}x^2 + \frac{1}{24}x^4 - \frac{1}{720}x^6 + \cdots\right)\left(2c_2 + 6c_3 x + 12c_4 x^2 + 20c_5 x^3 + 30c_6 x^4 + \cdots\right)$$

$$+ \sum_{n=0}^{\infty} c_n x^n$$

$$= \left[2c_2 + 6c_3 x + (12c_4 - c_2)x^2 + (20c_5 - 3c_3)x^3 + \left(30c_6 - 6c_4 + \frac{1}{12}c_2\right)x^4 + \cdots\right]$$

$$+ [c_0 + c_1 x + c_2 x^2 + c_3 x^3 + c_4 x^4 + \cdots]$$

$$= (c_0 + 2c_2) + (c_1 + 6c_3)x + 12c_4 x^2 + (20c_5 - 2c_3)x^3 + \left(30c_6 - 5c_4 + \frac{1}{12}c_2\right)x^4 + \cdots$$

$$= 0.$$

Thus

$$c_0 + 2c_2 = 0$$

$$c_1 + 6c_3 = 0$$

$$12c_4 = 0$$

$$20c_5 - 2c_3 = 0$$

$$30c_6 - 5c_4 + \frac{1}{12}c_2 = 0$$

and

$$c_2 = -\frac{1}{2}c_0$$

$$c_3 = -\frac{1}{6}c_1$$

82

$$c_4 = 0$$

$$c_5 = \frac{1}{10}c_3$$

$$c_6 = \frac{1}{6}c_4 - \frac{1}{360}c_0$$

Choosing $c_0 = 1$ and $c_1 = 0$ we find

$$c_2 = -\frac{1}{2}, \quad c_3 = 0, \quad c_4 = 0, \quad c_5 = 0, \quad c_6 = \frac{1}{720}$$

and so on. For $c_0 = 0$ and $c_1 = 1$ we find

$$c_2 = 0, \quad c_3 = -\frac{1}{6}, \quad c_4 = 0, \quad c_5 = -\frac{1}{60}, \quad c_6 = 0$$

and so on. Thus, two solutions are

$$y_1 = 1 - \frac{1}{2}x^2 + \frac{1}{720}x^6 + \cdots \quad \text{and} \quad y_2 = x - \frac{1}{6}x^3 - \frac{1}{60}x^5 + \cdots.$$

15. Substituting $y = \sum_{n=0}^{\infty} c_n x^{n+r}$ into the differential equation we obtain

$$2x^2 y'' + xy' - (x+1)y$$

$$= \left(2r^2 - r - 1\right)c_0 x^r + \sum_{k=1}^{\infty} [2(k+r)(k+r-1)c_k + (k+r)c_k - c_k - c_{k-1}]x^{k+r}$$

$$= 0$$

which implies $\quad\quad\quad 2r^2 - r - 1 = (2r+1)(r-1) = 0$

and $\quad\quad\quad [(k+r)(2k+2r-1) - 1]c_k - c_{k-1} = 0$

The indicial roots are $r = 1$ and $r = -1/2$. For $r = 1$ the recurrence relation is

$$c_k = \frac{c_{k-1}}{k(2k+3)}, \quad k = 1, 2, 3, \ldots,$$

so $\quad\quad\quad c_1 = \frac{1}{5}c_0, \quad c_2 = \frac{1}{70}c_0, \quad c_3 = \frac{1}{1,890}c_0.$

For $r = -1/2$ the recurrence relation is

$$c_k = \frac{c_{k-1}}{k(2k-3)}, \quad k = 1, 2, 3, \ldots,$$

so $\quad\quad\quad c_1 = -c_0, \quad c_2 = -\frac{1}{2}c_0, \quad c_3 = -\frac{1}{18}c_0.$

Two linearly independent solutions are

$$y_1 = C_1 x \left(1 + \frac{1}{5}x + \frac{1}{70}x^2 + \frac{1}{1,890}x^3 + \cdots\right)$$

and

$$y_2 = C_2 x^{-1/2}\left(1 - x - \frac{1}{2}x^2 - \frac{1}{18}x^3 - \cdots\right).$$

18. Substituting $y = \sum_{n=0}^{\infty} c_n x^{n+r}$ into the differential equation we obtain

$$x^2 y'' - xy' + \left(x^2 + 1\right) y = \left(r^2 - 2r + 1\right) c_0 x^r + r^2 c_1 x^{r+1}$$

$$+ \sum_{k=2}^{\infty} [(k+r)(k+r-1)c_k - (k+r)c_k + c_k + c_{k-2}]x^{k+r}$$

$$= 0$$

which implies
$$r^2 - 2r + 1 = (r-1)^2 = 0$$

$$r^2 c_1 = 0$$

$$[(k+r)(k+r-2) + 1]c_k + c_{k-2} = 0.$$

The indicial roots are $r_1 = r_2 = 1$, so $c_1 = 0$ and

$$c_k = -\frac{c_{k-2}}{k^2}, \quad k = 2, 3, 4, \ldots.$$

Thus
$$c_2 = -\frac{1}{4}c_0$$

$$c_3 = c_5 = c_7 = \cdots = 0$$

$$c_4 = \frac{1}{64}c_0$$

$$c_6 = -\frac{1}{2,304}c_0$$

and one solution is

$$y_1 = c_0 x\left(1 - \frac{1}{4}x^2 + \frac{1}{64}x^4 - \frac{1}{2,304}x^6 + \cdots\right).$$

A second solution is

$$y_2 = y_1 \int \frac{e^{dx/x}}{y_1^2}\, dx = y_1 \int \frac{x\, dx}{x^2\left(1 - \frac{1}{4}x^2 + \frac{1}{64}x^4 - \frac{1}{2304}x^6 + \cdots\right)^2}$$

$$= y_1 \int \frac{dx}{x\left(1 - \frac{1}{2}x^2 + \frac{3}{32}x^4 - \frac{5}{576}x^6 + \cdots\right)}$$

$$= y_1 \int \frac{1}{x}\left(1 + \frac{1}{2}x^2 + \frac{5}{32}x^4 + \frac{23}{576}x^6 + \cdots\right) dx$$

$$= y_1 \int \left(\frac{1}{x} + \frac{1}{2}x + \frac{5}{32}x^3 + \frac{23}{576}x^5 + \cdots \right) dx$$

$$= y_1 \ln x + y_1 \left(\frac{1}{4}x^2 + \frac{5}{128}x^4 + \frac{23}{3,456}x^6 + \cdots \right).$$

7 The Laplace Transform

Exercises 7.1

3. $\mathcal{L}\{f(t)\} = \displaystyle\int_0^1 te^{-st}dt + \int_1^\infty e^{-st}dt = \left(-\dfrac{1}{s}te^{-st} - \dfrac{1}{s^2}e^{-st}\right)\Big|_0^1 - \dfrac{1}{s}e^{-st}\Big|_1^\infty$

$\qquad = \left(-\dfrac{1}{s}e^{-s} - \dfrac{1}{s^2}e^{-s}\right) - \left(0 - \dfrac{1}{s^2}\right) - \dfrac{1}{s}(0 - e^{-s}) = \dfrac{1}{s^2}(1 - e^{-s}), \quad s > 0$

6. $\mathcal{L}\{f(t)\} = \displaystyle\int_{\pi/2}^\infty (\cos t)e^{-st}dt = \left(-\dfrac{s}{s^2+1}e^{-st}\cos t + \dfrac{1}{s^2+1}e^{-st}\sin t\right)\Big|_{\pi/2}^\infty$

$\qquad = 0 - \left(0 + \dfrac{1}{s^2+1}e^{-\pi s/2}\right) = -\dfrac{1}{s^2+1}e^{-\pi s/2}, \quad s > 0$

9. $f(t) = \begin{cases} 1 - t, & 0 < t < 1 \\ 0, & t > 0 \end{cases}$

$\qquad \mathcal{L}\{f(t)\} = \displaystyle\int_0^1 (1-t)e^{-st}\,dt = \left(-\dfrac{1}{s}(1-t)e^{-st} + \dfrac{1}{s^2}e^{-st}\right)\Big|_0^1 = \dfrac{1}{s^2}e^{-s} + \dfrac{1}{s} - \dfrac{1}{s^2}, \quad s > 0$

12. $\mathcal{L}\{f(t)\} = \displaystyle\int_0^\infty e^{-2t-5}e^{-st}dt = e^{-5}\int_0^\infty e^{-(s+2)t}dt = -\dfrac{e^{-5}}{s+2}e^{-(s+2)t}\Big|_0^\infty = \dfrac{e^{-5}}{s+2}, \quad s > -2$

15. $\mathcal{L}\{f(t)\} = \displaystyle\int_0^\infty e^{-t}(\sin t)e^{-st}dt = \int_0^\infty (\sin t)e^{-(s+1)t}dt$

$\qquad = \left(\dfrac{-(s+1)}{(s+1)^2+1}e^{-(s+1)t}\sin t - \dfrac{1}{(s+1)^2+1}e^{-(s+1)t}\cos t\right)\Big|_0^\infty$

$\qquad = \dfrac{1}{(s+1)^2+1} = \dfrac{1}{s^2+2s+2}, \quad s > -1$

18. $\mathcal{L}\{f(t)\} = \displaystyle\int_0^\infty t(\sin t)e^{-st}dt$

$\qquad = \left[\left(-\dfrac{t}{s^2+1} - \dfrac{2s}{(s^2+1)^2}\right)(\cos t)e^{-st} - \left(\dfrac{st}{s^2+1} + \dfrac{s^2-1}{(s^2+1)^2}\right)(\sin t)e^{-st}\right]_0^\infty$

$\qquad = \dfrac{2s}{(s^2+1)^2}, \quad s > 0$

21. $\mathcal{L}\{4t - 10\} = \dfrac{4}{s^2} - \dfrac{10}{s}$

24. $\mathcal{L}\{-4t^2 + 16t + 9\} = -4\dfrac{2}{s^3} + \dfrac{16}{s^2} + \dfrac{9}{s}$

27. $\mathcal{L}\{1 + e^{4t}\} = \dfrac{1}{s} + \dfrac{1}{s-4}$

30. $\mathcal{L}\{e^{2t} - 2 + e^{-2t}\} = \dfrac{1}{s-2} - \dfrac{2}{s} + \dfrac{1}{s+2}$

33. $\mathcal{L}\{\sinh kt\} = \dfrac{k}{s^2 - k^2}$

36. $\mathcal{L}\{e^{-t}\cosh t\} = \mathcal{L}\left\{e^{-t}\dfrac{e^t + e^{-t}}{2}\right\} = \mathcal{L}\left\{\dfrac{1}{2} + \dfrac{1}{2}e^{-2t}\right\} = \dfrac{1}{2s} + \dfrac{1}{2(s+2)}$

39. Let $u = st$ so that $du = s\,dt$ and $\mathcal{L}\{t^\alpha\} = \displaystyle\int_0^\infty e^{-st}t^\alpha dt = \int_0^\infty e^{-u}\left(\dfrac{u}{s}\right)^\alpha \dfrac{1}{s}\,du = \dfrac{1}{s^{\alpha+1}}\Gamma(\alpha+1)$

for $\alpha > -1$.

42. $\mathcal{L}\{t^{3/2}\} = \dfrac{\Gamma(5/2)}{s^{5/2}} = \dfrac{3\sqrt{\pi}}{4s^{5/2}}$

Exercises 7.2

3. $\mathcal{L}^{-1}\left\{\dfrac{1}{s^2} - \dfrac{48}{s^5}\right\} = \mathcal{L}^{-1}\left\{\dfrac{1}{s^2} - \dfrac{48}{24}\cdot\dfrac{4!}{s^5}\right\} = t - 2t^4$

6. $\mathcal{L}^{-1}\left\{\dfrac{(s+2)^2}{s^3}\right\} - \mathcal{L}^{-1}\left\{\dfrac{1}{s} + 4\cdot\dfrac{1}{s^2} + 2\cdot\dfrac{2}{s^3}\right\} = 1 + 4t + 2t^2$

9. $\mathcal{L}^{-1}\left\{\dfrac{1}{4s+1}\right\} - \mathcal{L}^{-1}\left\{\dfrac{1}{4}\cdot\dfrac{1}{s+1/4}\right\} = \dfrac{1}{4}e^{-t/4}$

12. $\mathcal{L}^{-1}\left\{\dfrac{10s}{s^2+16}\right\} = 10\cos 4t$

15. $\mathcal{L}^{-1}\left\{\dfrac{1}{s^2-16}\right\} - \mathcal{L}^{-1}\left\{\dfrac{1/8}{s-4} - \dfrac{1/8}{s+4}\right\} = \dfrac{1}{8}e^{4t} - \dfrac{1}{8}e^{-4t} = \dfrac{1}{4}\sinh 4t$

18. $\mathcal{L}^{-1}\left\{\dfrac{s+1}{s^2+2}\right\} = \mathcal{L}^{-1}\left\{\dfrac{s}{s^2+2} + \dfrac{1}{\sqrt{2}}\cdot\dfrac{\sqrt{2}}{s^2+2}\right\} = \cos\sqrt{2}\,t + \dfrac{1}{\sqrt{2}}\sin\sqrt{2}\,t$

21. $\mathcal{L}^{-1}\left\{\dfrac{s}{s^2+2s-3}\right\} = \mathcal{L}^{-1}\left\{\dfrac{1}{4}\cdot\dfrac{1}{s-1} + \dfrac{3}{4}\cdot\dfrac{1}{s+3}\right\} = \dfrac{1}{4}e^t + \dfrac{3}{4}e^{-3t}$

24. $\mathcal{L}^{-1}\left\{\dfrac{s-3}{(s-\sqrt{3})(s+\sqrt{3})}\right\} = \mathcal{L}^{-1}\left\{\dfrac{s}{s^2-3} - \sqrt{3}\cdot\dfrac{\sqrt{3}}{s^2-3}\right\} = \cosh\sqrt{3}\,t - \sqrt{3}\sinh\sqrt{3}\,t$

27. $\mathcal{L}^{-1}\left\{\dfrac{2s+4}{(s-2)(s^2+4s+3)}\right\} = \mathcal{L}^{-1}\left\{\dfrac{8}{15}\cdot\dfrac{1}{s-2} - \dfrac{1}{3}\cdot\dfrac{1}{s+1} - \dfrac{1}{5}\cdot\dfrac{1}{s+3}\right\} = \dfrac{8}{15}e^{2t} - \dfrac{1}{3}e^{-t} - \dfrac{1}{5}e^{-3t}$

30. $\mathscr{L}^{-1}\left\{\dfrac{s-1}{s^2(s^2+1)}\right\} = \mathscr{L}^{-1}\left\{\dfrac{1}{s} - \dfrac{1}{s^2} - \dfrac{s}{s^2+1} + \dfrac{1}{s^2+1}\right\} = 1 - t - \cos t + \sin t$

33. $\mathscr{L}^{-1}\left\{\dfrac{1}{(s^2+1)(s^2+4)}\right\} = \mathscr{L}^{-1}\left\{\dfrac{1}{3}\cdot\dfrac{1}{s^2+1} - \dfrac{1}{6}\cdot\dfrac{2}{s^2+4}\right\} = \dfrac{1}{3}\sin t - \dfrac{1}{6}\sin 2t$

Exercises 7.3

3. $\mathscr{L}\left\{t^3 e^{-2t}\right\} = \dfrac{3!}{(s+2)^4}$

6. $\mathscr{L}\left\{e^{-2t}\cos 4t\right\} = \dfrac{s+2}{(s+2)^2+16}$

9. $\mathscr{L}\left\{t\left(e^t + e^{2t}\right)^2\right\} = \mathscr{L}\left\{te^{2t} + 2te^{3t} + te^{4t}\right\} = \dfrac{1}{(s-2)^2} + \dfrac{2}{(s-3)^2} + \dfrac{1}{(s-4)^2}$

12. $\mathscr{L}\left\{e^t \cos^2 3t\right\} = \mathscr{L}\left\{\dfrac{1}{2}e^t + \dfrac{1}{2}e^t \cos 6t\right\} = \dfrac{1}{2}\dfrac{1}{s-1} + \dfrac{1}{2}\dfrac{s-1}{(s-1)^2+36}$

15. $\mathscr{L}^{-1}\left\{\dfrac{1}{s^2-6s+10}\right\} = \mathscr{L}^{-1}\left\{\dfrac{1}{(s-3)^2+1^2}\right\} = e^{3t}\sin t$

18. $\mathscr{L}^{-1}\left\{\dfrac{2s+5}{s^2+6s+34}\right\} = \mathscr{L}^{-1}\left\{2\dfrac{(s+3)}{(s+3)^2+5^2} - \dfrac{1}{5}\dfrac{5}{(s+3)^2+5^2}\right\} = 2e^{-3t}\cos 5t - \dfrac{1}{5}e^{-3t}\sin 5t$

21. $\mathscr{L}^{-1}\left\{\dfrac{2s-1}{s^2(s+1)^3}\right\} = \mathscr{L}^{-1}\left\{\dfrac{5}{s} - \dfrac{1}{s^2} - \dfrac{5}{s+1} - \dfrac{4}{(s+1)^2} - \dfrac{3}{2}\dfrac{2}{(s+1)^3}\right\} = 5 - t - 5e^{-t} - 4te^{-t} - \dfrac{3}{2}t^2 e^{-t}$

24. $\mathscr{L}\{e^{2-t}\,\mathcal{U}(t-2)\} = \mathscr{L}\left\{e^{-(t-2)}\,\mathcal{U}(t-2)\right\} = \dfrac{e^{-2s}}{s+1}$

27. $\mathscr{L}\{\cos 2t\,\mathcal{U}(t-\pi)\} = \mathscr{L}\{\cos 2(t-\pi)\,\mathcal{U}(t-\pi)\} = \dfrac{se^{-\pi s}}{s^2+4}$

30. $\mathscr{L}\left\{te^{t-5}\,\mathcal{U}(t-5)\right\} = \mathscr{L}\left\{(t-5)e^{t-5}\,\mathcal{U}(t-5) + 5e^{t-5}\,\mathcal{U}(t-5)\right\} = \dfrac{e^{-5s}}{(s-1)^2} + \dfrac{5e^{-5s}}{s-1}$

33. $\mathscr{L}^{-1}\left\{\dfrac{e^{-\pi s}}{s^2+1}\right\} = \sin(t-\pi)\,\mathcal{U}(t-\pi)$

36. $\mathscr{L}^{-1}\left\{\dfrac{e^{-2s}}{s^2(s-1)}\right\} = \mathscr{L}^{-1}\left\{-\dfrac{e^{-2s}}{s} - \dfrac{e^{-2s}}{s^2} + \dfrac{e^{-2s}}{s-1}\right\} = -\mathcal{U}(t-2) - (t-2)\,\mathcal{U}(t-2) + e^{t-2}\,\mathcal{U}(t-2)$

39. $\mathscr{L}\{t^2 \sinh t\} = \dfrac{d^2}{ds^2}\left(\dfrac{1}{s^2-1}\right) = \dfrac{6s^2+2}{(s^2-1)^3}$

42. $\mathcal{L}\left\{te^{-3t}\cos 3t\right\} = -\dfrac{d}{ds}\left(\dfrac{s+3}{(s+3)^2+9}\right) = \dfrac{(s+3)^2-9}{[(s+3)^2+9]^2}$

45. (c)

48. (b)

51. $\mathcal{L}\{2-4\mathcal{U}(t-3)\} = \dfrac{2}{s} - \dfrac{4}{s}e^{-3s}$

54. $\mathcal{L}\left\{\sin t\,\mathcal{U}\left(t-\dfrac{3\pi}{2}\right)\right\} = \mathcal{L}\left\{-\cos\left(t-\dfrac{3\pi}{2}\right)\mathcal{U}\left(t-\dfrac{3\pi}{2}\right)\right\} = -\dfrac{se^{-3\pi s/2}}{s^2+1}$

57. $\mathcal{L}\{f(t)\} = \mathcal{L}\{\mathcal{U}(t-a)-\mathcal{U}(t-b)\} = \dfrac{e^{-as}}{s} - \dfrac{e^{-bs}}{s}$

60. $\mathcal{L}^{-1}\left\{\dfrac{2}{s} - \dfrac{3e^{-s}}{s^2} + \dfrac{5e^{-2s}}{s^2}\right\} = 2 - 3(t-1)\mathcal{U}(t-1) + 5(t-2)\mathcal{U}(t-2)$

$$= \begin{cases} 2, & 0\le t<1 \\ -3t+5, & 1\le t<2 \\ 2t-5, & t\ge 2 \end{cases}$$

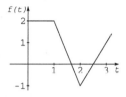

63. Since

$$t^2 - 3t = t(t-3) = (t-2+2)(t-2-1) = (t-2)^2 + (t-2) - 2$$

we have

$$\mathcal{L}\{(t^2-3t)\mathcal{U}(t-2)\} = \mathcal{L}\{(t-2)^2\mathcal{U}(t-2) + (t-2)\mathcal{U}(t-2) - 2\mathcal{U}(t-2)\}$$

$$= \dfrac{2}{s^3}e^{-2s} + \dfrac{1}{s^2}e^{-2s} - \dfrac{2}{s}e^{-2s}.$$

Using the alternative form of the second translation theorem we obtain

$$\mathcal{L}\{(t^2-3t)\mathcal{U}(t-2)\} = e^{-2s}\mathcal{L}\{(t+2)^2 - 3(t+2)\}$$

$$= e^{-2s}\mathcal{L}\{t^2+t-2\} = e^{-2s}\left(\dfrac{2}{s^3} + \dfrac{1}{s^2} - \dfrac{2}{s}\right).$$

Exercises 7.4

3. $\mathcal{L}\{y''+3y'\} = \mathcal{L}\{y''\} + 3\mathcal{L}\{y'\} = s^2Y(s) - sy(0) - y'(0) + 3[sY(s)-y(0)] = (s^2+3s)Y(s) - s - 2$

Exercises 7.4

6. We solve $\mathscr{L}\{y'' + y\} = \mathscr{L}\{1\} = 1/s$.

$$s^2 Y(s) - sy(0) - y'(0) + Y(s) = \frac{1}{s}$$

$$(s^2 + 1)Y(s) - 2s - 3 = \frac{1}{s}$$

$$Y(s) = \frac{1}{s(s^2 + 1)} + \frac{2s + 3}{s^2 + 1}$$

9. $\mathscr{L}\left\{\int_0^t e^{-\tau} \cos \tau \, d\tau\right\} = \frac{1}{s} \mathscr{L}\left\{e^{-t} \cos t\right\} = \frac{1}{s} \frac{s+1}{(s+1)^2 + 1} = \frac{s+1}{s(s^2 + 2s + 2)}$

12. $\mathscr{L}\left\{\int_0^t \sin \tau \cos(t - \tau) \, d\tau\right\} = \mathscr{L}\{\sin t\} \mathscr{L}\{\cos t\} = \frac{s}{(s^2 + 1)^2}$

15. $\mathscr{L}\{1 * t^3\} = \frac{1}{s} \frac{3!}{s^4} = \frac{6}{s^5}$

18. $\mathscr{L}\{t^2 * te^t\} = \dfrac{2}{s^3(s-1)^2}$

21. $\mathscr{L}^{-1}\left\{\dfrac{1}{s+5} F(s)\right\} = e^{-5t} * f(t) = \displaystyle\int_0^t f(\tau) e^{-5(t-\tau)} d\tau$

24. We use repeated applications of $\int_0^t f(\tau) \, d\tau = \mathscr{L}^{-1}\{F(s)/s\}$.

$$\mathscr{L}^{-1}\left\{\frac{1}{s(s-1)}\right\} = \int_0^t e^\tau \, d\tau = e^t - 1$$

$$\mathscr{L}^{-1}\left\{\frac{1}{s^2(s-1)}\right\} = \int_0^t (e^\tau - 1) \, d\tau = e^t - t - 1$$

$$\mathscr{L}^{-1}\left\{\frac{1}{s^3(s-1)}\right\} = \int_0^t (e^\tau - \tau - 1) \, d\tau = e^t - \frac{1}{2}t^2 - t - 1$$

27. $\mathscr{L}^{-1}\left\{\dfrac{s}{(s^2 + 4)^2}\right\} = \cos 2t * \dfrac{1}{2} \sin 2t = \dfrac{1}{2} \displaystyle\int_0^t \cos 2\tau \sin 2(t - \tau) \, d\tau$

$$= \frac{1}{2} \int_0^t \cos 2\tau (\sin 2t \cos 2\tau - \cos 2t \sin 2\tau) \, d\tau = \frac{1}{2}\left[\sin 2t \int_0^t \cos^2 2\tau \, d\tau - \cos 2t \int_0^t \frac{1}{2} \sin 4\tau \, d\tau\right]$$

$$= \frac{1}{2} \sin 2t \left[\frac{1}{2}\tau + \frac{1}{8} \sin 4\tau\right]_0^t - \frac{1}{4} \cos 2t \left[-\frac{1}{4} \cos 4\tau\right]_0^t$$

$$= \frac{1}{2} \sin 2t \left(\frac{1}{2}t + \frac{1}{8} \sin 4t\right) + \frac{1}{16} \cos 2t (\cos 4t - 1)$$

$$= \frac{1}{4}t \sin 2t + \frac{1}{16} \sin 2t \sin 4t + \frac{1}{16} \cos 2t \cos 4t - \frac{1}{16} \cos 2t$$

$$= \frac{1}{4}t \sin 2t + \frac{1}{16}\left[\sin 2t(2 \sin 2t \cos 2t) + \cos 2t \left(\cos^2 2t - \sin^2 2t\right) - \cos 2t\right]$$

$$= \frac{1}{4}t \sin 2t + \frac{1}{16}\cos 2t \left[2 \sin^2 2t + \cos^2 2t - \sin^2 2t - 1\right] = \frac{1}{4}t \sin 2t$$

30. $f * (g + h) = \displaystyle\int_0^t f(\tau)[g(t - \tau) + h(t - \tau)]\, d\tau = \int_0^t f(\tau)g(t - \tau)\, d\tau + \int_0^t f(\tau)h(t - \tau)\, d\tau$

$$- \int_0^t f(\tau)[g(t \quad \tau) \mid h(t \quad \tau)]\, d\tau - f * y + f * h$$

33. $\mathcal{L}\{f(t)\} = \dfrac{1}{1 - e^{-bs}}\displaystyle\int_0^b \frac{a}{b}te^{-st}dt = \frac{a}{s}\left(\frac{1}{bs} - \frac{1}{e^{bs} - 1}\right)$

36. $\mathcal{L}\{f(t)\} = \dfrac{1}{1 - e^{-2\pi s}}\displaystyle\int_0^{\pi} e^{-st}\sin t\, dt = \frac{1}{s^2 + 1} \cdot \frac{1}{1 - e^{-\pi s}}$

Exercises 7.5

3. The Laplace transform of the differential equation is

$$s\mathcal{L}\{y\} - y(0) + 4\mathcal{L}\{y\} = \frac{1}{s + 4}.$$

Solving for $\mathcal{L}\{y\}$ we obtain $\mathcal{L}\{y\} = \dfrac{1}{(s + 4)^2} + \dfrac{2}{s + 4}.$

Thus $\qquad\qquad\qquad\qquad y = te^{-4t} + 2e^{-4t}.$

6. The Laplace transform of the differential equation is

$$s^2\mathcal{L}\{y\} - sy(0) - y'(0) - 6\left[s\mathcal{L}\{y\} - y(0)\right] + 13\mathcal{L}\{y\} = 0.$$

Solving for $\mathcal{L}\{y\}$ we obtain

$$\mathcal{L}\{y\} = -\frac{3}{s^2 - 6s + 13} = -\frac{3}{2}\frac{2}{(s - 3)^2 + 2^2}.$$

Thus $\qquad\qquad\qquad\qquad y = -\dfrac{3}{2}e^{3t}\sin 2t.$

9. The Laplace transform of the differential equation is

$$s^2\mathcal{L}\{y\} - sy(0) - y'(0) - 4\left[s\mathcal{L}\{y\} - y(0)\right] + 4\mathcal{L}\{y\} = \frac{6}{(s - 2)^4}.$$

Solving for $\mathcal{L}\{y\}$ we obtain $\mathcal{L}\{y\} = \dfrac{1}{20}\dfrac{5!}{(s - 2)^6}.$ Thus, $y = \dfrac{1}{20}t^5 e^{2t}.$

Exercises 7.5

12. The Laplace transform of the differential equation is

$$s^2 \mathcal{L}\{y\} - sy(0) - y'(0) + 16\,\mathcal{L}\{y\} = \frac{1}{s}.$$

Solving for $\mathcal{L}\{y\}$ we obtain

$$\mathcal{L}\{y\} = \frac{s^2 + 2s + 1}{s(s^2 + 16)} = \frac{1}{16}\frac{1}{s} + \frac{15}{16}\frac{s}{s^2 + 4^2} + \frac{1}{2}\frac{4}{s^2 + 4^2}.$$

Thus
$$y = \frac{1}{16} + \frac{15}{16}\cos 4t + \frac{1}{2}\sin 4t.$$

15. The Laplace transform of the differential equation is

$$2\left[s^3 \mathcal{L}\{y\} - s^2(0) - sy'(0) - y''(0)\right] + 3[s^2 \mathcal{L}\{y\} - sy(0) - y'(0)] - 3[s\mathcal{L}\{y\} - y(0)] - 2\,\mathcal{L}\{y\} = \frac{1}{s+1}.$$

Solving for $\mathcal{L}\{y\}$ we obtain

$$\mathcal{L}\{y\} = \frac{2s + 3}{(s+1)(s-1)(2s+1)(s+2)} = \frac{1}{2}\frac{1}{s+1} + \frac{5}{18}\frac{1}{s-1} - \frac{8}{9}\frac{1}{s+1/2} + \frac{1}{9}\frac{1}{s+2}.$$

Thus
$$y = \frac{1}{2}e^{-t} + \frac{5}{18}e^{t} - \frac{8}{9}e^{-t/2} + \frac{1}{9}e^{-2t}.$$

18. The Laplace transform of the differential equation is

$$s^4 \mathcal{L}\{y\} - s^3 y(0) - s^2 y'(0) - sy''(0) - y'''(0) - \mathcal{L}\{y\} = \frac{1}{s^2}.$$

Solving for $\mathcal{L}\{y\}$ we obtain

$$\mathcal{L}\{y\} = \frac{1}{s^2(s^4 - 1)} = -\frac{1}{s^2} + \frac{1}{4}\frac{1}{s-1} - \frac{1}{4}\frac{1}{s+1} + \frac{1}{2}\frac{1}{s^2 + 1}.$$

Thus
$$y = -t + \frac{1}{4}e^{t} - \frac{1}{4}e^{-t} + \frac{1}{2}\sin t.$$

21. The Laplace transform of the differential equation is

$$s\mathcal{L}\{y\} - y(0) + 2\,\mathcal{L}\{y\} = \frac{1}{s^2} - e^{-s}\frac{s+1}{s^2}.$$

Solving for $\mathcal{L}\{y\}$ we obtain

$$\mathcal{L}\{y\} = \frac{1}{s^2(s+2)} - e^{-s}\frac{s+1}{s^2(s+1)} = -\frac{1}{4}\frac{1}{s} + \frac{1}{2}\frac{1}{s^2} + \frac{1}{4}\frac{1}{s+2} - e^{-s}\left[\frac{1}{4}\frac{1}{s} + \frac{1}{2}\frac{1}{s^2} - \frac{1}{4}\frac{1}{s+2}\right].$$

Thus
$$y = -\frac{1}{4} + \frac{1}{2}t + \frac{1}{4}e^{-2t} - \left[\frac{1}{4} + \frac{1}{2}(t-1) - \frac{1}{4}e^{-2(t-1)}\right]\mathscr{U}(t-1).$$

24. The Laplace transform of the differential equation is

$$s^2 \mathcal{L}\{y\} - sy(0) - y'(0) - 5\,[s\mathcal{L}\{y\} - y(0)] + 6\,\mathcal{L}\{y\} = \frac{e^{-s}}{s}.$$

Solving for $\mathcal{L}\{y\}$ we obtain

$$\mathcal{L}\{y\} = e^{-s}\frac{1}{s(s-2)(s-3)} + \frac{1}{(s-2)(s-3)}$$

$$= e^{-s}\left[\frac{1}{6}\frac{1}{s} - \frac{1}{2}\frac{1}{s-2} + \frac{1}{3}\frac{1}{s-3}\right] - \frac{1}{s-2} + \frac{1}{s-3}.$$

Thus

$$y = \left[\frac{1}{6} - \frac{1}{2}e^{2(t-1)} + \frac{1}{3}e^{3(t-1)}\right]\mathcal{U}(t-1) + e^{3t} - e^{2t}.$$

27. Taking the Laplace transform of both sides of the differential equation and letting $c = y(0)$ we obtain

$$\mathcal{L}\{y''\} + \mathcal{L}\{2y'\} + \mathcal{L}\{y\} = 0$$

$$s^2\mathcal{L}\{y\} - sy(0) - y'(0) + 2s\,\mathcal{L}\{y\} - 2y(0) + \mathcal{L}\{y\} = 0$$

$$s^2\mathcal{L}\{y\} - cs - 2 + 2s\,\mathcal{L}\{y\} - 2c + \mathcal{L}\{y\} = 0$$

$$\left(s^2 + 2s + 1\right)\mathcal{L}\{y\} = cs + 2c + 2$$

$$\mathcal{L}\{y\} = \frac{cs}{(s+1)^2} + \frac{2c+2}{(s+1)^2}$$

$$= c\frac{s+1-1}{(s+1)^2} + \frac{2c+2}{(s+1)^2}$$

$$= \frac{c}{s+1} + \frac{c+2}{(s+1)^2}.$$

Therefore,

$$y(t) = c\mathcal{L}^{-1}\left\{\frac{1}{s+1}\right\} + (c+2)\,\mathcal{L}^{-1}\left\{\frac{1}{(s+1)^2}\right\} = ce^{-t} + (c+2)te^{-t}.$$

To find c we let $y(1) = 2$. Then $2 = ce^{-1} + (c+2)e^{-1} = 2(c+1)e^{-1}$ and $c = e - 1$.

Thus

$$y(t) = (e-1)e^{-t} + (e+1)te^{-t}.$$

30. The Laplace transform of the given equation is

$$\mathcal{L}\{f\} = \mathcal{L}\{2t\} - 4\mathcal{L}\{\sin t\}\mathcal{L}\{f\}.$$

Solving for $\mathcal{L}\{f\}$ we obtain

$$\mathcal{L}\{f\} = \frac{2s^2 + 2}{s^2(s^2 + 5)} = \frac{2}{5}\frac{1}{s^2} + \frac{8}{5\sqrt{5}}\frac{\sqrt{5}}{s^2 + 5}.$$

Thus

$$f(t) = \frac{2}{5}t + \frac{8}{5\sqrt{5}}\sin\sqrt{5}\,t.$$

93

33. The Laplace transform of the given equation is

$$\mathscr{L}\{f\} + \mathscr{L}\{1\}\mathscr{L}\{f\} = \mathscr{L}\{1\}.$$

Solving for $\mathscr{L}\{f\}$ we obtain $\mathscr{L}\{f\} = \dfrac{1}{s+1}$. Thus, $f(t) = e^{-t}$.

36. The Laplace transform of the given equation is

$$\mathscr{L}\{t\} - 2\mathscr{L}\{f\} = \mathscr{L}\{e^t - e^{-t}\}\mathscr{L}\{f\}.$$

Solving for $\mathscr{L}\{f\}$ we obtain

$$\mathscr{L}\{f\} = \frac{s^2 - 1}{2s^4} = \frac{1}{2}\frac{1}{s^2} - \frac{1}{12}\frac{3!}{s^4}.$$

Thus

$$f(t) = \frac{1}{2}t - \frac{1}{12}t^3.$$

39. From equation (3) in the text the differential equation is

$$0.005\frac{di}{dt} + i + 50\int_0^t i(\tau)\,d\tau = 100[1 - \mathscr{U}(t-1)], \quad i(0) = 0.$$

The Laplace transform of this equation is

$$0.005[s\,\mathscr{L}\{i\} - i(0)] + \mathscr{L}\{i\} + 50\frac{1}{s}\mathscr{L}\{i\} = 100\left[\frac{1}{s} - \frac{1}{s}e^{-s}\right].$$

Solving for $\mathscr{L}\{i\}$ we obtain

$$\mathscr{L}\{i\} = \frac{20{,}000}{(s+100)^2}(1 - e^{-s}).$$

Thus

$$i(t) = 20{,}000te^{-100t} - 20{,}000(t-1)e^{-100(t-1)}\,\mathscr{U}(t-1).$$

42. The differential equation is

$$10\frac{dq}{dt} + 10q = 30e^t - 30e^t\,\mathscr{U}(t-1.5).$$

The Laplace transform of this equation is

$$s\mathscr{L}\{q\} - q_0 + \mathscr{L}\{q\} = \frac{3}{s-1} - \frac{3e^{1.5}}{s-1.5}e^{-1.5s}.$$

Solving for $\mathscr{L}\{q\}$ we obtain

$$\mathscr{L}\{q\} = \left(q_0 - \frac{3}{2}\right)\cdot\frac{1}{s+1} + \frac{3}{2}\cdot\frac{1}{s-1}3e^{1.5}\left(\frac{-2/5}{s+1} + \frac{2/5}{s-1.5}\right)e^{-1.55}.$$

Thus

$$q(t) = \left(q_0 - \frac{3}{2}\right)e^{-t} + \frac{3}{2}e^t + \frac{6}{5}e^{1.5}\left(e^{-(t-1.5)} - e^{1.5(t-1.5)}\right)\mathscr{U}(t-1.5).$$

45. (a) The differential equation is

$$\frac{di}{dt} + 10i = \sin t + \cos\left(t - \frac{3\pi}{2}\right)\mathscr{U}\left(t - \frac{3\pi}{2}\right), \quad i(0) = 0.$$

The Laplace transform of this equation is

$$s\mathscr{L}\{i\} + 10\mathscr{L}\{i\} = \frac{1}{s^2 + 1} + \frac{se^{-3\pi s/2}}{s^2 + 1}.$$

Solving for $\mathscr{L}\{i\}$ we obtain

$$\mathscr{L}\{i\} = \frac{1}{(s^2 + 1)(s + 10)} + \frac{s}{(s^2 + 1)(s + 10)}e^{-3\pi s/2}$$

$$= \frac{1}{101}\left(\frac{1}{s + 10} - \frac{s}{s^2 + 1} + \frac{10}{s^2 + 1}\right) + \frac{1}{101}\left(\frac{-10}{s + 10} + \frac{10s}{s^2 + 1} + \frac{1}{s^2 + 1}\right)e^{-3\pi s/2}.$$

Thus

$$i(t) = \frac{1}{101}\left(e^{-10t} - \cos t + 10\sin t\right)$$

$$+ \frac{1}{101}\left[-10e^{-10(t-3\pi/2)} + 10\cos\left(t - \frac{3\pi}{2}\right) + \sin\left(t - \frac{3\pi}{2}\right)\right]\mathscr{U}\left(t - \frac{3\pi}{2}\right).$$

(b)

The maximum value of $i(t)$ is approximately 0.1 at $t = 1.7$, the minimum is approximately -0.1 at 4.7.

48. The differential equation is

$$\frac{d^2q}{dt^2} + 20\frac{dq}{dt} + 200q = 150, \quad q(0) = q'(0) = 0.$$

The Laplace transform of this equation is

$$s^2\mathscr{L}\{q\} + 20s\mathscr{L}\{q\} + 200\mathscr{L}\{q\} = \frac{150}{s}.$$

Solving for $\mathscr{L}\{q\}$ we obtain

$$\mathscr{L}\{q\} = \frac{150}{s(s^2 + 20s + 200)} = \frac{3}{4}\frac{1}{s} - \frac{3}{4}\frac{s + 10}{(s + 10)^2 + 10^2} - \frac{3}{4}\frac{10}{(s + 10)^2 + 10^2}.$$

Thus

$$q(t) = \frac{3}{4} - \frac{3}{4}e^{-10t}\cos 10t - \frac{3}{4}e^{-10t}\sin 10t$$

and

$$i(t) = q'(t) = 15e^{-10t}\sin 10t.$$

If $E(t) = 150 - 150\,\mathcal{U}(t-2)$, then

$$\mathcal{L}\{q\} = \frac{150}{s(s^2 + 20s + 200)}\left(1 - e^{-2s}\right)$$

$$q(t) = \frac{3}{4} - \frac{3}{4}e^{-10t}\cos 10t - \frac{3}{4}e^{-10t}\sin 10t - \left[\frac{3}{4} - \frac{3}{4}e^{-10(t-2)}\cos 10(t-2)\right.$$

$$\left. - \frac{3}{4}e^{-10(t-2)}\sin 10(t-2)\right]\mathcal{U}(t-2).$$

51. The differential equation is

$$\frac{d^2q}{dt^2} + \frac{1}{LC}q = \frac{E_0}{L}e^{-kt}, \quad q(0) = q'(0) = 0.$$

The Laplace transform of this equation is

$$s^2\mathcal{L}\{q\} + \frac{1}{LC}\mathcal{L}\{q\} = \frac{E_0}{L}\frac{1}{s+k}.$$

Solving for $\mathcal{L}\{q\}$ we obtain

$$\mathcal{L}\{q\} = \frac{E_0}{L}\frac{1}{(s+k)(s^2 + 1/LC)} = \frac{E_0}{L}\left(\frac{1/(k^2 + 1/LC)}{s+k} - \frac{s/(k^2 + 1/LC)}{s^2 + 1/LC} + \frac{k/(k^2 + 1/LC)}{s^2 + 1/LC}\right).$$

Thus $\qquad q(t) = \frac{E_0}{L(k^2 + 1/LC)}\left[e^{-kt} - \cos\left(t/\sqrt{LC}\right) + k\sqrt{LC}\,\sin\left(t/\sqrt{LC}\right)\right].$

54. Recall from Chapter 5 that $mx'' = -kx + f(t)$. Now $m = W/g = 16/32 = 1/2$ slug, and $k = 4.5$, so the differential equation is

$$\frac{1}{2}x'' + 4.5x = 4\sin 3t + 2\cos 3t \quad \text{or} \quad x'' + 9x = 8\sin 3t + 4\cos 3t.$$

The initial conditions are $x(0) = x'(0) = 0$. The Laplace transform of the differential equation is

$$s^2\mathcal{L}\{x\} + 9\mathcal{L}\{x\} = \frac{24}{s^2 + 9} + \frac{4s}{s^2 + 9}.$$

Solving for $\mathcal{L}\{x\}$ we obtain

$$\mathcal{L}\{x\} = \frac{4s + 24}{(s^2 + 9)^2} = \frac{2}{3}\frac{2(3)s}{(s^2 + 9)^2} + \frac{12}{27}\frac{2(3)^3}{(s^2 + 9)^2}.$$

Thus $\qquad x(t) = \frac{2}{3}t\sin 3t + \frac{4}{9}(\sin 3t - 3t\cos 3t) = \frac{2}{3}t\sin 3t + \frac{4}{9}\sin 3t - \frac{4}{3}t\cos 3t.$

57. The differential equation is

$$EI\frac{d^4y}{dx^4} = \frac{2w_0}{L}\left[\frac{L}{2} - x + \left(x - \frac{L}{2}\right)\mathcal{U}\left(x - \frac{L}{2}\right)\right].$$

Taking the Laplace transform of both sides and using $y(0) = y'(0) = 0$ we obtain

$$s^4 \mathcal{L}\{y\} - sy''(0) - y'''(0) = \frac{2w_0}{EIL}\left[\frac{L}{2s} - \frac{1}{s^2} + \frac{1}{s^2}e^{-Ls/2}\right].$$

Letting $y''(0) = c_1$ and $y'''(0) = c_2$ we have

$$\mathcal{L}\{y\} = \frac{c_1}{s^3} + \frac{c_2}{s^4} + \frac{2w_0}{EIL}\left[\frac{L}{2s^5} - \frac{1}{s^6} + \frac{1}{s^6}e^{-Ls/2}\right]$$

so that

$$y(x) = \frac{1}{2}c_1x^2 + \frac{1}{6}c_2x^3 + \frac{2w_0}{EIL}\left[\frac{L}{48}x^4 - \frac{1}{120}x^5 + \frac{1}{120}\left(x - \frac{L}{2}\right)\mathcal{U}\left(x - \frac{L}{2}\right)\right]$$

$$= \frac{1}{2}c_1x^2 + \frac{1}{6}c_2x^3 + \frac{w_0}{60EIL}\left[\frac{5L}{2}x^4 - x^5 + \left(x - \frac{L}{2}\right)^5\mathcal{U}\left(x - \frac{L}{2}\right)\right].$$

To find c_1 and c_2 we compute

$$y''(x) = c_1 + c_2x + \frac{w_0}{60EIL}\left[30Lx^2 - 20x^3 + 20\left(x - \frac{L}{2}\right)^3\mathcal{U}\left(x - \frac{L}{2}\right)\right]$$

and

$$y'''(x) = c_2 + \frac{w_0}{60EIL}\left[60Lx - 60x^2 + 60\left(x - \frac{L}{2}\right)^2\mathcal{U}\left(x - \frac{L}{2}\right)\right].$$

Then $y''(L) = y'''(L) = 0$ yields the system

$$c_1 + c_2L + \frac{w_0}{60EIL}\left[30L^3 - 20L^3 + \frac{5}{2}L^3\right] = c_1 + c_2L + \frac{5w_0L^2}{24EI} = 0$$

$$c_2 + \frac{w_0}{60EIL}[60L^2 - 60L^2 + 15L^2] = c_2 + \frac{w_0L}{4EI} = 0.$$

Solving for c_1 and c_2 we obtain $c_1 = w_0L^2/24EI$ and $c_2 = -w_0L/4EI$. Thus

$$y(x) = \frac{w_0L^2}{48EI}x^2 - \frac{w_0L}{24EI} + \frac{w_0}{60EIL}\left[\frac{5L}{2}x^4 - x^5 + \left(x - \frac{L}{2}\right)^5\mathcal{U}\left(x - \frac{L}{2}\right)\right].$$

Exercises 7.6

3. The Laplace transform of the differential equation yields

$$\mathcal{L}\{y\} = \frac{1}{s^2 + 1}\left(1 + e^{-2\pi s}\right)$$

so that

$$y = \sin t + \sin t\, \mathcal{U}(t - 2\pi).$$

6. The Laplace transform of the differential equation yields

$$\mathcal{L}\{y\} = \frac{s}{s^2 + 1} + \frac{1}{s^2 + 1}(e^{-2\pi s} + e^{-4\pi s})$$

so that
$$y = \cos t + \sin t[\,\mathcal{U}(t - 2\pi) + \mathcal{U}(t - 4\pi)\,].$$

9. The Laplace transform of the differential equation yields

$$\mathcal{L}\{y\} = \frac{1}{(s+2)^2 + 1}e^{-2\pi s}$$

so that
$$y = e^{-2(t-2\pi)}\sin t\,\mathcal{U}(t - 2\pi).$$

12. The Laplace transform of the differential equation yields

$$\mathcal{L}\{y\} = \frac{1}{(s-1)^2(s-6)} + \frac{e^{-2s} + e^{-4s}}{(s-1)(s-6)}$$

$$= -\frac{1}{25}\frac{1}{s-1} - \frac{1}{5}\frac{1}{(s-1)^2} + \frac{1}{25}\frac{1}{s-6} + \left[-\frac{1}{5}\frac{1}{s-1} + \frac{1}{5}\frac{1}{s-6}\right]\left(e^{-2s} + e^{-4s}\right)$$

so that
$$y = -\frac{1}{25}e^t - \frac{1}{5}te^t + \frac{1}{25}e^{6t} + \left[-\frac{1}{5}e^{t-2} + \frac{1}{5}e^{6(t-2)}\right]\mathcal{U}(t-2)$$

$$+ \left[-\frac{1}{5}e^{t-4} + \frac{1}{5}e^{6(t-4)}\right]\mathcal{U}(t-4).$$

Exercises 7.7

3. Taking the Laplace transform of the system gives

$$s\mathcal{L}\{x\} + 1 = \mathcal{L}\{x\} - 2\mathcal{L}\{y\}$$
$$s\mathcal{L}\{y\} - 2 = 5\mathcal{L}\{x\} - \mathcal{L}\{y\}$$

so that
$$\mathcal{L}\{x\} = \frac{-s-5}{s^2+9} = -\frac{s}{s^2+9} - \frac{5}{3}\frac{3}{s^2+9}$$

and
$$x = -\cos 3t - \frac{5}{3}\sin 3t.$$

Then
$$y = \frac{1}{2}x - \frac{1}{2}x' = 2\cos 3t - \frac{7}{3}\sin 3t.$$

6. Taking the Laplace transform of the system gives

$$(s+1)\,\mathcal{L}\{x\} - (s-1)\mathcal{L}\{y\} = -1$$
$$s\,\mathcal{L}\{x\} + (s+2)\,\mathcal{L}\{y\} = 1$$

98

so that
$$\mathcal{L}\{y\} = \frac{s+1/2}{s^2+s+1} = \frac{s+1/2}{(s+1/2)^2+(\sqrt{3}/2)^2}$$

and
$$\mathcal{L}\{x\} = \frac{-3/2}{s^2+s+1} = \frac{-3/2}{(s+1/2)^2+(\sqrt{3}/2)^2}.$$

Then
$$y = e^{-t/2}\cos\frac{\sqrt{3}}{2}t \quad \text{and} \quad x = e^{-t/2}\sin\frac{\sqrt{3}}{2}t.$$

9. Adding the equations and then subtracting them gives

$$\frac{d^2x}{dt^2} = \frac{1}{2}t^2 + 2t$$

$$\frac{d^2y}{dt^2} = \frac{1}{2}t^2 - 2t.$$

Taking the Laplace transform of the system gives

and
$$\mathcal{L}\{x\} = 8\frac{1}{s} + \frac{1}{24}\frac{4!}{s^5} + \frac{1}{3}\frac{3!}{s^4}$$

$$\mathcal{L}\{y\} = \frac{1}{24}\frac{4!}{s^5} - \frac{1}{3}\frac{3!}{s^4}$$

so that
$$x = 8 + \frac{1}{24}t^4 + \frac{1}{3}t^3 \quad \text{and} \quad y = \frac{1}{24}t^4 - \frac{1}{3}t^3.$$

12. Taking the Laplace transform of the system gives

$$(s-4)\mathcal{L}\{x\} + 2\mathcal{L}\{y\} = \frac{2e^{-s}}{s}$$

$$-3\mathcal{L}\{x\} + (s+1)\mathcal{L}\{y\} = \frac{1}{2} + \frac{e^{-s}}{s}$$

so that
$$\mathcal{L}\{x\} = \frac{-1/2}{(s-1)(s-2)} + e^{-s}\frac{1}{(s-1)(s-2)}$$

$$= \left[\frac{1}{2}\frac{1}{s-1} - \frac{1}{2}\frac{1}{s-2}\right] + e^{-s}\left[-\frac{1}{s-1} + \frac{1}{s-2}\right]$$

and
$$\mathcal{L}\{y\} = \frac{e^{-s}}{s} + \frac{s/4-1}{(s-1)(s-2)} + e^{-s}\frac{-s/2+2}{(s-1)(s-2)}$$

$$= \frac{3}{4}\frac{1}{s-1} - \frac{1}{2}\frac{1}{s-2} + e^{-s}\left[\frac{1}{s} - \frac{3}{2}\frac{1}{s-1} + \frac{1}{s-2}\right].$$

Then
$$x = \frac{1}{2}e^t - \frac{1}{2}e^{2t} + \left[-e^{t-1} + e^{2(t-1)}\right]\mathscr{U}(t-1)$$

and
$$y = \frac{3}{4}e^t - \frac{1}{2}e^{2t} + \left[1 - \frac{3}{2}e^{t-1} + e^{2(t-1)}\right]\mathscr{U}(t-1).$$

15. (a) By Kirchoff's first law we have $i_1 = i_2 + i_3$. By Kirchoff's second law, on each loop we have $E(t) = Ri_1 + L_1i_2'$ and $E(t) = Ri_1 + L_2i_3'$ or $L_1i_2' + Ri_2 + Ri_3 = E(t)$ and $L_2i_3' + Ri_2 + Ri_3 = E(t)$.

(b) Taking the Laplace transform of the system
$$0.01i_2' + 5i_2 + 5i_3 = 100$$

$$0.0125i_3' + 5i_2 + 5i_3 = 100$$

gives
$$(s + 500)\,\mathscr{L}\{i_2\} + 500\mathscr{L}\{i_3\} = \frac{10,000}{s}$$

$$400\mathscr{L}\{i_2\} + (s + 400)\,\mathscr{L}\{i_3\} = \frac{8,000}{s}$$

so that
$$\mathscr{L}\{i_3\} = \frac{8,000}{s^2 + 900s} = \frac{80}{9}\frac{1}{s} - \frac{80}{9}\frac{1}{s + 900}.$$

Then
$$i_3 = \frac{80}{9} - \frac{80}{9}e^{-900t} \quad \text{and} \quad i_2 = 20 - 0.0025i_3' - i_3 = \frac{100}{9} - \frac{100}{9}e^{-900t}.$$

(c) $i_1 = i_2 + i_3 = 20 - 20e^{-900t}$

18. Taking the Laplace transform of the system
$$0.5i_1' + 50i_2 = 60$$

$$0.005i_2' + i_2 - i_1 = 0$$

gives
$$s\,\mathscr{L}\{i_1\} + 100\,\mathscr{L}\{i_2\} = \frac{120}{s}$$

$$-200\,\mathscr{L}\{i_1\} + (s + 200)\,\mathscr{L}\{i_2\} = 0$$

so that
$$\mathscr{L}\{i_2\} = \frac{24,000}{s(s^2 + 200s + 20,000)} = \frac{6}{5}\frac{1}{s} - \frac{6}{5}\frac{s + 100}{(s + 100)^2 + 100^2} - \frac{6}{5}\frac{100}{(s + 100)^2 + 100^2}.$$

Then
$$i_2 = \frac{6}{5} - \frac{6}{5}e^{-100t}\cos 100t - \frac{6}{5}e^{-100t}\sin 100t$$

and
$$i_1 = 0.005i_2' + i_2 = \frac{6}{5} - \frac{6}{5}e^{-100t}\cos 100t.$$

21. Taking the Laplace transform of the system

$$4\theta_1'' + \theta_2'' + 8\theta_1 = 0$$

$$\theta_1'' + \theta_2'' + 2\theta_2 = 0$$

gives

$$4\left(s^2 + 2\right)\mathcal{L}\{\theta_1\} + s^2\mathcal{L}\{\theta_2\} = 3s$$

$$s^2\mathcal{L}\{\theta_1\} + \left(s^2 + 2\right)\mathcal{L}\{\theta_2\} = 0$$

so that

$$\left(3s^2 + 4\right)\left(s^2 + 4\right)\mathcal{L}\{\theta_2\} = -3s^3$$

or

$$\mathcal{L}\{\theta_2\} = \frac{1}{2}\frac{s}{s^2 + 4/3} - \frac{3}{2}\frac{s}{s^2 + 4}.$$

Then

$$\theta_2 = \frac{1}{2}\cos\frac{2}{\sqrt{3}}t - \frac{3}{2}\cos 2t \quad \text{and} \quad \theta_1'' = -\theta_2'' - 2\theta_2$$

so that

$$\theta_1 = \frac{1}{4}\cos\frac{2}{\sqrt{3}}t + \frac{3}{4}\cos 2t.$$

──────── **Chapter 7 Review Exercises** ────────

3. False; consider $f(t) = t^{-1/2}$.

6. False; consider $f(t) = 1$ and $g(t) = 1$.

9. $\mathcal{L}\{\sin 2t\} = \dfrac{2}{s^2 + 4}$

12. $\mathcal{L}\{\sin 2t\,\mathcal{U}(t - \pi)\} = \mathcal{L}\{\sin 2(t - \pi)\mathcal{U}(t - \pi)\} = \dfrac{2}{s^2 + 4}e^{-\pi s}$

15. $\mathcal{L}^{-1}\left\{\dfrac{1}{(s - 5)^3}\right\} = \mathcal{L}^{-1}\left\{\dfrac{1}{2}\dfrac{2}{(s - 5)^3}\right\} = \dfrac{1}{2}t^2 e^{5t}$

18. $\mathcal{L}^{-1}\left\{\dfrac{1}{s^2}e^{-5s}\right\} = (t - 5)\mathcal{U}(t - 5)$

21. $\mathcal{L}\{e^{-5t}\}$ exists for $s > -5$.

24. $1 * 1 = \displaystyle\int_0^t d\tau = t$

27. **(a)** $f(t) = 2 - 2\mathcal{U}(t - 2) + [(t - 2) + 2]\mathcal{U}(t - 2) = 2 + (t - 2)\mathcal{U}(t - 2)$

(b) $\mathscr{L}\{f(t)\} = \dfrac{2}{s} + \dfrac{1}{s^2}e^{-2s}$

(c) $\mathscr{L}\{e^t f(t)\} = \dfrac{2}{s-1} + \dfrac{1}{(s-1)^2}e^{-2(s-1)}$

30. Taking the Laplace transform of the differential equation we obtain

$$\mathscr{L}\{y\} = \frac{1}{(s-1)^2(s^2 - 8s + 20)}$$

$$= \frac{6}{169}\frac{1}{s-1} + \frac{1}{13}\frac{1}{(s-1)^2} - \frac{6}{169}\frac{s-4}{(s-4)^2 + 2^2} + \frac{5}{338}\frac{2}{(s-4)^2 + 2^2}$$

so that
$$y = \frac{6}{169}e^t + \frac{1}{13}te^t - \frac{6}{169}e^{4t}\cos 2t + \frac{5}{338}e^{4t}\sin 2t.$$

33. Taking the Laplace transform of the differential equation we obtain

$$\mathscr{L}\{y\} = \frac{s^3 + 2}{s^3(s-5)} - \frac{2 + 2s + s^2}{s^3(s-5)}e^{-s}$$

$$= -\frac{2}{125}\frac{1}{s} - \frac{2}{25}\frac{1}{s^2} - \frac{1}{5}\frac{2}{s^3} + \frac{127}{125}\frac{1}{s-5} - \left[-\frac{37}{125}\frac{1}{s} - \frac{12}{25}\frac{1}{s^2} - \frac{1}{5}\frac{2}{s^3} + \frac{37}{125}\frac{1}{s-5}\right]e^{-s}$$

so that

$$y = -\frac{2}{125} - \frac{2}{25}t - \frac{1}{5}t^2 + \frac{127}{125}e^{5t} - \left[-\frac{37}{125} - \frac{12}{25}(t-1) - \frac{1}{5}(t-1)^2 + \frac{37}{125}e^{5(t-1)}\right]\mathscr{U}(t-1).$$

36. Taking the Laplace transform of the integral equation we obtain

$$(\mathscr{L}\{f\})^2 = 6 \cdot \frac{6}{s^4} \quad \text{or} \quad \mathscr{L}\{f\} = \pm 6 \cdot \frac{1}{s^2}$$

so that $f(t) = \pm 6t$.

39. The integral equation is

$$10i + 2\int_0^t i(\tau)\,d\tau = 2t^2 + 2t.$$

Taking the Laplace transform we obtain

$$\mathscr{L}\{i\} = \left(\frac{4}{s^3} + \frac{2}{s^2}\right)\frac{s}{10s + 2} = \frac{s+2}{s^2(5s+2)} = -\frac{9}{s} + \frac{2}{s^2} + \frac{45}{5s+1} = -\frac{9}{s} + \frac{2}{s^2} + \frac{9}{s+1/5}.$$

Thus
$$i(t) = -9 + 2t + 9e^{-t/5}.$$

42. Taking the Laplace transform of the given differential equation we obtain

$$\mathscr{L}\{y\} = \frac{c_1}{2} \cdot \frac{2s}{s^4 + 4} + \frac{c_2}{4} \cdot \frac{4}{s^4 + 4} + \frac{w_0}{4EI} \cdot \frac{4}{s^4 + 4}e^{-s\pi/2}$$

so that

$$y = \frac{c_1}{2} \sin x \sinh x + \frac{c_2}{4}(\sin x \cosh x - \cos x \sinh x)$$

$$+ \frac{w_0}{4EI}\left[\sin\left(x - \frac{\pi}{2}\right)\cosh\left(x - \frac{\pi}{2}\right) - \cos\left(x - \frac{\pi}{2}\right)\sinh\left(x - \frac{\pi}{2}\right)\right]\mathcal{U}\left(x - \frac{\pi}{2}\right)$$

where $y''(0) = c_1$ and $y'''(0) = c_2$. Using $y(\pi) = 0$ and $y'(\pi) = 0$ we find

$$c_1 = \frac{w_0}{EI}\frac{\sinh\frac{\pi}{2}}{\sinh\pi}, \qquad c_2 = -\frac{w_0}{EI}\frac{\cosh\frac{\pi}{2}}{\sinh\pi}.$$

Hence

$$y = \frac{w_0}{2EI}\frac{\sinh\frac{\pi}{2}}{\sinh\pi}\sin x \sinh x - \frac{w_0}{4EI}\frac{\cosh\frac{\pi}{2}}{\sinh\pi}(\sin x \cosh x - \cos x \sinh x)$$

$$+ \frac{w_0}{4EI}\left[\sin\left(x - \frac{\pi}{2}\right)\cosh\left(x - \frac{\pi}{2}\right) - \cos\left(x - \frac{\pi}{2}\right)\sinh\left(x - \frac{\pi}{2}\right)\right]\mathcal{U}\left(x - \frac{\pi}{2}\right).$$

103

8 Systems of Linear First-Order Differential Equations

_____ **Exercises 8.1** _____

3. Let $\mathbf{X} = \begin{pmatrix} x \\ y \\ z \end{pmatrix}$. Then

$$\mathbf{X}' = \begin{pmatrix} -3 & 4 & -9 \\ 6 & -1 & 0 \\ 10 & 4 & 3 \end{pmatrix} \mathbf{X}.$$

6. Let $\mathbf{X} = \begin{pmatrix} x \\ y \end{pmatrix}$. Then

$$\mathbf{X}' = \begin{pmatrix} -3 & 4 \\ 5 & 9 \end{pmatrix} \mathbf{X} + \begin{pmatrix} e^{-t} \sin 2t \\ 4e^{-t} \cos 2t \end{pmatrix}.$$

9. $\dfrac{dx}{dt} = x - y + 2z + e^{-t} - 3t; \quad \dfrac{dy}{dt} = 3x - 4y + z + 2e^{-t} + t; \quad \dfrac{dz}{dt} = -2x + 5y + 6z + 2e^{-t} - t$

12. Since

$$\mathbf{X}' = \begin{pmatrix} 5\cos t - 5\sin t \\ 2\cos t - 4\sin t \end{pmatrix} e^t \quad \text{and} \quad \begin{pmatrix} -2 & 5 \\ -2 & 4 \end{pmatrix} \mathbf{X} = \begin{pmatrix} 5\cos t - 5\sin t \\ 2\cos t - 4\sin t \end{pmatrix} e^t$$

we see that

$$\mathbf{X}' = \begin{pmatrix} -2 & 5 \\ -2 & 4 \end{pmatrix} \mathbf{X}.$$

15. Since

$$\mathbf{X}' = \begin{pmatrix} 0 \\ 0 \\ 0 \end{pmatrix} \quad \text{and} \quad \begin{pmatrix} 1 & 2 & 1 \\ 6 & -1 & 0 \\ -1 & -2 & -1 \end{pmatrix} \mathbf{X} = \begin{pmatrix} 0 \\ 0 \\ 0 \end{pmatrix}$$

we see that

$$\mathbf{X}' = \begin{pmatrix} 1 & 2 & 1 \\ 6 & -1 & 0 \\ -1 & -2 & -1 \end{pmatrix} \mathbf{X}.$$

18. Yes, since $W(\mathbf{X}_1, \mathbf{X}_2) = 8e^{2t} \neq 0$ and $\mathbf{X}_1$ and $\mathbf{X}_2$ are linearly independent on $-\infty < t < \infty$.

21. Since

$$\mathbf{X}'_p = \begin{pmatrix} 2 \\ -1 \end{pmatrix} \quad \text{and} \quad \begin{pmatrix} 1 & 4 \\ 3 & 2 \end{pmatrix} \mathbf{X}_p + \begin{pmatrix} 2 \\ -4 \end{pmatrix} t + \begin{pmatrix} -7 \\ -18 \end{pmatrix} = \begin{pmatrix} 2 \\ -1 \end{pmatrix}$$

we see that
$$\mathbf{X}'_p = \begin{pmatrix} 1 & 4 \\ 3 & 2 \end{pmatrix} \mathbf{X}_p + \begin{pmatrix} 2 \\ -4 \end{pmatrix} t + \begin{pmatrix} -7 \\ -18 \end{pmatrix}.$$

24. Since
$$\mathbf{X}_p = \begin{pmatrix} 3\cos 3t \\ 0 \\ -3\sin 3t \end{pmatrix} \quad \text{and} \quad \begin{pmatrix} 1 & 2 & 3 \\ -4 & 2 & 0 \\ -6 & 1 & 0 \end{pmatrix} \mathbf{X}_p + \begin{pmatrix} -1 \\ 4 \\ 3 \end{pmatrix} \sin 3t = \begin{pmatrix} 3\cos 3t \\ 0 \\ -3\sin 3t \end{pmatrix}$$

we see that
$$\mathbf{X}'_p = \begin{pmatrix} 1 & 2 & 3 \\ -4 & 2 & 0 \\ -6 & 1 & 0 \end{pmatrix} \mathbf{X}_p + \begin{pmatrix} -1 \\ 4 \\ 3 \end{pmatrix} \sin 3t.$$

———— **Exercises 8.2** ————

3. The system is
$$\mathbf{X}' = \begin{pmatrix} -4 & 2 \\ -5/2 & 2 \end{pmatrix} \mathbf{X}$$

and $\det(\mathbf{A} - \lambda\mathbf{I}) = (\lambda - 1)(\lambda + 3) = 0$. For $\lambda_1 = 1$ we obtain
$$\begin{pmatrix} -5 & 2 & | & 0 \\ -5/2 & 1 & | & 0 \end{pmatrix} \Longrightarrow \begin{pmatrix} -5 & 2 & | & 0 \\ 0 & 0 & | & 0 \end{pmatrix} \quad \text{so that} \quad \mathbf{K}_1 = \begin{pmatrix} 2 \\ 5 \end{pmatrix}.$$

For $\lambda_2 = -3$ we obtain
$$\begin{pmatrix} -1 & 2 & | & 0 \\ -5/2 & 5 & | & 0 \end{pmatrix} \Longrightarrow \begin{pmatrix} -1 & 2 & | & 0 \\ 0 & 0 & | & 0 \end{pmatrix} \quad \text{so that} \quad \mathbf{K}_2 = \begin{pmatrix} 2 \\ 1 \end{pmatrix}.$$

Then
$$\mathbf{X} = c_1 \begin{pmatrix} 2 \\ 5 \end{pmatrix} e^t + c_2 \begin{pmatrix} 2 \\ 1 \end{pmatrix} e^{-3t}.$$

6. The system is
$$\mathbf{X}' = \begin{pmatrix} -6 & 2 \\ -3 & 1 \end{pmatrix} \mathbf{X}$$

and $\det(\mathbf{A} - \lambda\mathbf{I}) = \lambda(\lambda + 5) = 0$. For $\lambda_1 = 0$ we obtain
$$\begin{pmatrix} -6 & 2 & | & 0 \\ -3 & 1 & | & 0 \end{pmatrix} \Longrightarrow \begin{pmatrix} 1 & -1/3 & | & 0 \\ 0 & 0 & | & 0 \end{pmatrix} \quad \text{so that} \quad \mathbf{K}_1 = \begin{pmatrix} 1 \\ 3 \end{pmatrix}.$$

For $\lambda_2 = -5$ we obtain
$$\begin{pmatrix} -1 & 2 & | & 0 \\ -3 & 6 & | & 0 \end{pmatrix} \Longrightarrow \begin{pmatrix} 1 & -2 & | & 0 \\ 0 & 0 & | & 0 \end{pmatrix} \quad \text{so that} \quad \mathbf{K}_2 = \begin{pmatrix} 2 \\ 1 \end{pmatrix}.$$

Then
$$\mathbf{X} = c_1 \begin{pmatrix} 1 \\ 3 \end{pmatrix} + c_2 \begin{pmatrix} 2 \\ 1 \end{pmatrix} e^{-5t}.$$

9. We have $\det(\mathbf{A} - \lambda\mathbf{I}) = -(\lambda+1)(\lambda-3)(\lambda+2) = 0$. For $\lambda_1 = -1$, $\lambda_2 = 3$, and $\lambda_3 = -2$ we obtain

$$\mathbf{K}_1 = \begin{pmatrix} -1 \\ 0 \\ 1 \end{pmatrix}, \quad \mathbf{K}_2 = \begin{pmatrix} 1 \\ 4 \\ 3 \end{pmatrix}, \quad \text{and} \quad \mathbf{K}_3 = \begin{pmatrix} 1 \\ -1 \\ 3 \end{pmatrix},$$

so that
$$\mathbf{X} = c_1 \begin{pmatrix} -1 \\ 0 \\ 1 \end{pmatrix} e^{-t} + c_2 \begin{pmatrix} 1 \\ 4 \\ 3 \end{pmatrix} e^{3t} + c_3 \begin{pmatrix} 1 \\ -1 \\ 3 \end{pmatrix} e^{-2t}.$$

12. We have $\det(\mathbf{A} - \lambda\mathbf{I}) = (\lambda - 3)(\lambda + 5)(6 - \lambda) = 0$. For $\lambda_1 = 3$, $\lambda_2 = -5$, and $\lambda_3 = 6$ we obtain

$$\mathbf{K}_1 = \begin{pmatrix} 1 \\ 1 \\ 0 \end{pmatrix}, \quad \mathbf{K}_2 = \begin{pmatrix} 1 \\ -1 \\ 0 \end{pmatrix}, \quad \text{and} \quad \mathbf{K}_3 = \begin{pmatrix} 2 \\ -2 \\ 11 \end{pmatrix},$$

so that
$$\mathbf{X} = c_1 \begin{pmatrix} 1 \\ 1 \\ 0 \end{pmatrix} e^{3t} + c_2 \begin{pmatrix} 1 \\ -1 \\ 0 \end{pmatrix} e^{-5t} + c_3 \begin{pmatrix} 2 \\ -2 \\ 11 \end{pmatrix} e^{6t}.$$

15. $\mathbf{X} = c_1 \begin{pmatrix} 0.382175 \\ 0.851161 \\ 0.359815 \end{pmatrix} e^{8.58979t} + c_2 \begin{pmatrix} 0.405188 \\ -0.676043 \\ 0.615458 \end{pmatrix} e^{2.25684t} + c_3 \begin{pmatrix} -0.923562 \\ -0.132174 \\ 0.35995 \end{pmatrix} e^{-0.0466321t}$

18. We have $\det(\mathbf{A} - \lambda\mathbf{I}) = (\lambda+1)^2 = 0$. For $\lambda_1 = -1$ we obtain

$$\mathbf{K} = \begin{pmatrix} 1 \\ 1 \end{pmatrix}.$$

A solution of $(\mathbf{A} - \lambda_1\mathbf{I})\mathbf{P} = \mathbf{K}$ is

$$\mathbf{P} = \begin{pmatrix} 0 \\ 1/5 \end{pmatrix}$$

so that
$$\mathbf{X} = c_1 \begin{pmatrix} 1 \\ 1 \end{pmatrix} e^{-t} + c_2 \left[\begin{pmatrix} 1 \\ 1 \end{pmatrix} t e^{-t} + \begin{pmatrix} 0 \\ 1/5 \end{pmatrix} e^{-t} \right].$$

21. We have $\det(\mathbf{A} - \lambda\mathbf{I}) = (1 - \lambda)(\lambda - 2)^2 = 0$. For $\lambda_1 = 1$ we obtain

$$\mathbf{K}_1 = \begin{pmatrix} 1 \\ 1 \\ 1 \end{pmatrix}.$$

ID#: SI55409
CARTON: 1 of 1
PO#:
INVOICE: 1225431358660

VIA: U2

PRIME-INDUCT

PAGE 1 OF 1
BATCH: 0513395

053/100

LOCATION	QTY	ISBN	AUTHOR/TITLE
K-16B-005-24	1	0-534-95574-6	ZILL 1ST CASE DIFF EQU 6ED
K-29E-014-13	1	0-534-95575-4	ZILL/WRIGHT/WRIGHT *SSM-DIFF EQU W/BVP
K-35F-014-09	1	0-534-95578-9	ZILL/WRIGHT/WRIGHT SSM-1ST CASE DIFF EQ 6E

WAREHOUSE INSTRUCTIONS

SLA: 7   BOX: staple

SALES SUPPORT

Date	Account	Contact
12/08/97	587540	10

SHIP TO: Yvonne like Yaz
Centenary College
Dept of Mathematics
2911 Centenary Blvd
P O Box 41188
Shreveport LA 711341188

ITP HIGHER EDUCATION

DISTRIBUTION CENTER
7625 EMPIRE DRIVE
FLORENCE, KY 41042

The enclosed materials are sent to you for your review by
MICHAEL W. WORLS

For $\lambda_2 = 2$ we obtain

$$\mathbf{K}_2 = \begin{pmatrix} 1 \\ 0 \\ 1 \end{pmatrix} \quad \text{and} \quad \mathbf{K}_3 = \begin{pmatrix} 1 \\ 1 \\ 0 \end{pmatrix}.$$

Then

$$\mathbf{X} = c_1 \begin{pmatrix} 1 \\ 1 \\ 1 \end{pmatrix} e^t + c_2 \begin{pmatrix} 1 \\ 0 \\ 1 \end{pmatrix} e^{2t} + c_3 \begin{pmatrix} 1 \\ 1 \\ 0 \end{pmatrix} e^{2t}.$$

24. We have $\det(\mathbf{A} - \lambda\mathbf{I}) = (1 - \lambda)(\lambda - 2)^2 = 0$. For $\lambda_1 = 1$ we obtain

$$\mathbf{K}_1 = \begin{pmatrix} 1 \\ 0 \\ 0 \end{pmatrix}.$$

For $\lambda_2 = 2$ we obtain

$$\mathbf{K} = \begin{pmatrix} 0 \\ -1 \\ 1 \end{pmatrix}.$$

A solution of $(\mathbf{A} - \lambda_2\mathbf{I})\mathbf{P} = \mathbf{K}$ is

$$\mathbf{P} = \begin{pmatrix} 0 \\ -1 \\ 0 \end{pmatrix}$$

so that

$$\mathbf{X} = c_1 \begin{pmatrix} 1 \\ 0 \\ 0 \end{pmatrix} e^t + c_2 \begin{pmatrix} 0 \\ -1 \\ 1 \end{pmatrix} e^{2t} + c_3 \left[\begin{pmatrix} 0 \\ -1 \\ 1 \end{pmatrix} te^{2t} + \begin{pmatrix} 0 \\ -1 \\ 0 \end{pmatrix} e^{2t} \right].$$

27. We have $\det(\mathbf{A} - \lambda\mathbf{I}) = (\lambda - 4)^2 = 0$. For $\lambda_1 = 4$ we obtain

$$\mathbf{K} = \begin{pmatrix} 2 \\ 1 \end{pmatrix}.$$

A solution of $(\mathbf{A} - \lambda_1\mathbf{I})\mathbf{P} = \mathbf{K}$ is

$$\mathbf{P} = \begin{pmatrix} 1 \\ 1 \end{pmatrix}$$

so that

$$\mathbf{X} = c_1 \begin{pmatrix} 2 \\ 1 \end{pmatrix} e^{4t} + c_2 \left[\begin{pmatrix} 2 \\ 1 \end{pmatrix} te^{4t} + \begin{pmatrix} 1 \\ 1 \end{pmatrix} e^{4t} \right].$$

If

$$\mathbf{X}(0) = \begin{pmatrix} -1 \\ 6 \end{pmatrix}$$

107

then $c_1 = -7$ and $c_2 = 13$.

33. We have $\det(\mathbf{A} - \lambda\mathbf{I}) = \lambda^2 - 8\lambda + 17 = 0$. For $\lambda_1 = 4 + i$ we obtain

$$\mathbf{K}_1 = \begin{pmatrix} -1 - i \\ 2 \end{pmatrix}$$

so that
$$\mathbf{X}_1 = \begin{pmatrix} -1 - i \\ 2 \end{pmatrix} e^{(4+i)t} = \begin{pmatrix} \sin t - \cos t \\ 2\cos t \end{pmatrix} e^{4t} + i \begin{pmatrix} -\sin t - \cos t \\ 2\sin t \end{pmatrix} e^{4t}.$$

Then
$$\mathbf{X} = c_1 \begin{pmatrix} \sin t - \cos t \\ 2\cos t \end{pmatrix} e^{4t} + c_2 \begin{pmatrix} -\sin t - \cos t \\ 2\sin t \end{pmatrix} e^{4t}.$$

36. We have $\det(\mathbf{A} - \lambda\mathbf{I}) = \lambda^2 + 2\lambda + 5 = 0$. For $\lambda_1 = -1 + 2i$ we obtain

$$\mathbf{K}_1 = \begin{pmatrix} 2 + 2i \\ 1 \end{pmatrix}$$

so that
$$\mathbf{X}_1 = \begin{pmatrix} 2 + 2i \\ 1 \end{pmatrix} e^{(-1+2i)t}$$

$$= \left(2\cos 2t - 2\sin 2t \cos 2t \right) e^{-t} + i \begin{pmatrix} 2\cos 2t + 2\sin 2t \\ \sin 2t \end{pmatrix} e^{-t}.$$

Then
$$\mathbf{X} = c_1 \begin{pmatrix} 2\cos 2t - 2\sin 2t \\ \cos 2t \end{pmatrix} e^{-t} + c_2 \begin{pmatrix} 2\cos 2t + 2\sin 2t \\ \sin 2t \end{pmatrix} e^{-t}.$$

39. We have $\det(\mathbf{A} - \lambda\mathbf{I}) = (1 - \lambda)(\lambda^2 - 2\lambda + 2) = 0$. For $\lambda_1 = 1$ we obtain

$$\mathbf{K}_1 = \begin{pmatrix} 0 \\ 2 \\ 1 \end{pmatrix}.$$

For $\lambda_2 = 1 + i$ we obtain

$$\mathbf{K}_2 = \begin{pmatrix} 1 \\ i \\ i \end{pmatrix}$$

so that
$$\mathbf{X}_2 = \begin{pmatrix} 1 \\ i \\ i \end{pmatrix} e^{(1+i)t} = \begin{pmatrix} \cos t \\ -\sin t \\ -\sin t \end{pmatrix} e^t + i \begin{pmatrix} \sin t \\ \cos t \\ \cos t \end{pmatrix} e^t.$$

Then
$$\mathbf{X} = c_1 \begin{pmatrix} 0 \\ 2 \\ 1 \end{pmatrix} e^t + c_2 \begin{pmatrix} \cos t \\ -\sin t \\ -\sin t \end{pmatrix} e^t + c_3 \begin{pmatrix} \sin t \\ \cos t \\ \cos t \end{pmatrix} e^t.$$

42. We have $\det(\mathbf{A} - \lambda\mathbf{I}) = -(\lambda + 2)(\lambda^2 + 4) = 0$. For $\lambda_1 = -2$ we obtain

$$\mathbf{K}_1 = \begin{pmatrix} 0 \\ -1 \\ 1 \end{pmatrix}.$$

For $\lambda_2 = 2i$ we obtain

$$\mathbf{K}_2 = \begin{pmatrix} -2 - 2i \\ 1 \\ 1 \end{pmatrix}$$

so that

$$\mathbf{X}_2 = \begin{pmatrix} -2 - 2i \\ 1 \\ 1 \end{pmatrix} e^{2it} = \begin{pmatrix} -2\cos 2t + 2\sin 2t \\ \cos 2t \\ \cos 2t \end{pmatrix} + i \begin{pmatrix} -2\cos 2t - 2\sin 2t \\ \sin 2t \\ \sin 2t \end{pmatrix}.$$

Then

$$\mathbf{X} = c_1 \begin{pmatrix} 0 \\ -1 \\ 1 \end{pmatrix} e^{-2t} + c_2 \begin{pmatrix} -2\cos 2t + 2\sin 2t \\ \cos 2t \\ \cos 2t \end{pmatrix} + c_3 \begin{pmatrix} -2\cos 2t - 2\sin 2t \\ \sin 2t \\ \sin 2t \end{pmatrix}.$$

—————— **Exercises 8.3** ——————————————————

3. From

$$\mathbf{X}' = \begin{pmatrix} 3 & -5 \\ 3/4 & -1 \end{pmatrix} \mathbf{X} + \begin{pmatrix} 1 \\ -1 \end{pmatrix} e^{t/2}$$

we obtain

$$\mathbf{X}_c = c_1 \begin{pmatrix} 10 \\ 3 \end{pmatrix} e^{3t/2} + c_2 \begin{pmatrix} 2 \\ 1 \end{pmatrix} e^{t/2}.$$

Then

$$\mathbf{\Phi} = \begin{pmatrix} 10e^{3t/2} & 2e^{t/2} \\ 3e^{3t/2} & e^{t/2} \end{pmatrix} \quad \text{and} \quad \mathbf{\Phi}^{-1} = \begin{pmatrix} \frac{1}{4}e^{-3t/2} & -\frac{1}{2}e^{-3t/2} \\ -\frac{3}{4}e^{-t/2} & \frac{5}{2}e^{-t/2} \end{pmatrix}$$

so that

$$\mathbf{U} = \int \mathbf{\Phi}^{-1}\mathbf{F}\, dt = \int \begin{pmatrix} \frac{3}{4}e^{-t} \\ -\frac{13}{4} \end{pmatrix} dt = \begin{pmatrix} -\frac{3}{4}e^{-t} \\ -\frac{13}{4}t \end{pmatrix}$$

and

$$\mathbf{X}_p = \mathbf{\Phi}\mathbf{U} = \begin{pmatrix} -13/2 \\ -13/4 \end{pmatrix} te^{t/2} + \begin{pmatrix} -15/2 \\ -9/4 \end{pmatrix} e^{t/2}.$$

6. From

$$\mathbf{X}' = \begin{pmatrix} 0 & 2 \\ -1 & 3 \end{pmatrix} \mathbf{X} + \begin{pmatrix} 2 \\ e^{-3t} \end{pmatrix}$$

we obtain

$$\mathbf{X}_c = c_1 \begin{pmatrix} 2 \\ 1 \end{pmatrix} e^t + c_2 \begin{pmatrix} 1 \\ 1 \end{pmatrix} e^{2t}.$$

Then
$$\Phi = \begin{pmatrix} 2e^t & e^{2t} \\ e^t & e^{2t} \end{pmatrix} \quad \text{and} \quad \Phi^{-1} = \begin{pmatrix} e^{-t} & -e^{-t} \\ -e^{-2t} & 2e^{-2t} \end{pmatrix}$$

so that
$$U = \int \Phi^{-1} F \, dt = \int \begin{pmatrix} 2e^{-t} - e^{-4t} \\ -2e^{-2t} + 2e^{-5t} \end{pmatrix} dt = \begin{pmatrix} -2e^{-t} + \frac{1}{4}e^{-4t} \\ e^{-2t} - \frac{2}{5}e^{-5t} \end{pmatrix}$$

and
$$X_p = \Phi U = \begin{pmatrix} \frac{1}{10}e^{-3t} - 3 \\ -\frac{3}{20}e^{-3t} - 1 \end{pmatrix}.$$

9. From
$$X' = \begin{pmatrix} 3 & 2 \\ -2 & -1 \end{pmatrix} X + \begin{pmatrix} 2 \\ 1 \end{pmatrix} e^{-t}$$

we obtain
$$X_c = c_1 \begin{pmatrix} 1 \\ -1 \end{pmatrix} e^t + c_2 \left[\begin{pmatrix} 1 \\ -1 \end{pmatrix} te^t + \begin{pmatrix} 0 \\ 1/2 \end{pmatrix} e^t \right].$$

Then
$$\Phi = \begin{pmatrix} e^t & te^t \\ -e^t & \frac{1}{2}e^t - te^t \end{pmatrix} \quad \text{and} \quad \Phi^{-1} = \begin{pmatrix} e^{-t} - 2te^{-t} & -2te^{-t} \\ 2e^{-t} & 2e^{-t} \end{pmatrix}$$

so that
$$U = \int \Phi^{-1} F \, dt = \int \begin{pmatrix} 2e^{-2t} - 6te^{-2t} \\ 6e^{-2t} \end{pmatrix} dt = \begin{pmatrix} \frac{1}{2}e^{-2t} + 3te^{-2t} \\ -3e^{-2t} \end{pmatrix}$$

and
$$X_p = \Phi U = \begin{pmatrix} 1/2 \\ -2 \end{pmatrix} e^{-t}.$$

12. From
$$X' = \begin{pmatrix} 1 & -1 \\ 1 & 1 \end{pmatrix} X + \begin{pmatrix} 3 \\ 3 \end{pmatrix} e^t$$

we obtain
$$X_c = c_1 \begin{pmatrix} -\sin t \\ \cos t \end{pmatrix} e^t + c_2 \begin{pmatrix} \cos t \\ \sin t \end{pmatrix} e^t.$$

Then
$$\Phi = \begin{pmatrix} -\sin t & \cos t \\ \cos t & \sin t \end{pmatrix} e^t \quad \text{and} \quad \Phi^{-1} = \begin{pmatrix} -\sin t & \cos t \\ \cos t & \sin t \end{pmatrix} e^{-t}$$

so that
$$U = \int \Phi^{-1} F \, dt = \int \begin{pmatrix} -3\sin t + 3\cos t \\ 3\cos t + 3\sin t \end{pmatrix} dt = \begin{pmatrix} 3\cos t + 3\sin t \\ 3\sin t - 3\cos t \end{pmatrix}$$

and
$$X_p = \Phi U = \begin{pmatrix} -3 \\ 3 \end{pmatrix} e^t.$$

15. From
$$X' = \begin{pmatrix} 0 & 1 \\ -1 & 0 \end{pmatrix} X + \begin{pmatrix} 0 \\ \sec t \tan t \end{pmatrix}$$

we obtain
$$\mathbf{X}_c = c_1 \begin{pmatrix} \cos t \\ -\sin t \end{pmatrix} + c_2 \begin{pmatrix} \sin t \\ \cos t \end{pmatrix}.$$

Then
$$\boldsymbol{\Phi} = \begin{pmatrix} \cos t & \sin t \\ \sin t & \cos t \end{pmatrix} t \quad \text{and} \quad \boldsymbol{\Phi}^{-1} = \begin{pmatrix} \cos t & -\sin t \\ \sin t & \cos t \end{pmatrix}$$

so that
$$\mathbf{U} = \int \boldsymbol{\Phi}^{-1}\mathbf{F}\, dt = \int \begin{pmatrix} -\tan^2 t \\ \tan t \end{pmatrix} dt = \begin{pmatrix} t - \tan t \\ \ln|\sec t| \end{pmatrix}$$

and
$$\mathbf{X}_p = \boldsymbol{\Phi}\mathbf{U} = \begin{pmatrix} \cos t \\ -\sin t \end{pmatrix} t + \begin{pmatrix} -\sin t \\ \sin t \tan t \end{pmatrix} + \begin{pmatrix} \sin t \\ \cos t \end{pmatrix} \ln|\sec t|.$$

18. From
$$\mathbf{X}' = \begin{pmatrix} 1 & -2 \\ 1 & -1 \end{pmatrix} \mathbf{X} + \begin{pmatrix} \tan t \\ 1 \end{pmatrix}$$

we obtain
$$\mathbf{X}_c = c_1 \begin{pmatrix} \cos t - \sin t \\ \cos t \end{pmatrix} + c_2 \begin{pmatrix} \cos t + \sin t \\ \sin t \end{pmatrix}.$$

Then
$$\boldsymbol{\Phi} = \begin{pmatrix} \cos t - \sin t & \cos t + \sin t \\ \cos t & \sin t \end{pmatrix} \quad \text{and} \quad \boldsymbol{\Phi}^{-1} = \begin{pmatrix} -\sin t & \cos t + \sin t \\ \cos t & \sin t - \cos t \end{pmatrix}$$

so that
$$\mathbf{U} = \int \boldsymbol{\Phi}^{-1}\mathbf{F}\, dt = \int \begin{pmatrix} 2\cos t + \sin t - \sec t \\ 2\sin t - \cos t \end{pmatrix} dt = \begin{pmatrix} 2\sin t - \cos t - \ln|\sec t + \tan t| \\ -2\cos t - \sin t \end{pmatrix}$$

and
$$\mathbf{X}_p = \boldsymbol{\Phi}\mathbf{U} = \begin{pmatrix} 3\sin t \cos t - \cos^2 t - 2\sin^2 t + (\sin t - \cos t)\ln|\sec t + \tan t| \\ \sin^2 t - \cos^2 t - \cos t(\ln|\sec t + \tan t|) \end{pmatrix}.$$

21. From
$$\mathbf{X}' = \begin{pmatrix} 3 & -1 \\ -1 & 3 \end{pmatrix} \mathbf{X} + \begin{pmatrix} 4e^{2t} \\ 4e^{4t} \end{pmatrix}$$

we obtain
$$\boldsymbol{\Phi} = \begin{pmatrix} -e^{4t} & e^{2t} \\ e^{4t} & e^{2t} \end{pmatrix}, \quad \boldsymbol{\Phi}^{-1} = \begin{pmatrix} -\frac{1}{2}e^{-4t} & \frac{1}{2}e^{-4t} \\ \frac{1}{2}e^{-2t} & \frac{1}{2}e^{-2t} \end{pmatrix},$$

and
$$\mathbf{X} = \boldsymbol{\Phi}\boldsymbol{\Phi}^{-1}(0)\mathbf{X}(0) + \boldsymbol{\Phi}\int_0^t \boldsymbol{\Phi}^{-1}\mathbf{F}\, ds = \boldsymbol{\Phi}\cdot\begin{pmatrix} 0 \\ 1 \end{pmatrix} + \boldsymbol{\Phi}\cdot\begin{pmatrix} e^{-2t} + 2t - 1 \\ e^{2t} + 2t - 1 \end{pmatrix}$$

$$= \begin{pmatrix} 2 \\ 2 \end{pmatrix} te^{2t} + \begin{pmatrix} -1 \\ 1 \end{pmatrix} e^{2t} + \begin{pmatrix} -2 \\ 2 \end{pmatrix} te^{4t} + \begin{pmatrix} 2 \\ 0 \end{pmatrix} e^{4t}.$$

24. (a) The eigenvalues are 0, 1, 3, and 4, with corresponding eigenvectors

$$\begin{pmatrix} -6 \\ -4 \\ 1 \\ 2 \end{pmatrix}, \quad \begin{pmatrix} 2 \\ 1 \\ 0 \\ 0 \end{pmatrix}, \quad \begin{pmatrix} 3 \\ 1 \\ 2 \\ 1 \end{pmatrix}, \quad \text{and} \quad \begin{pmatrix} -1 \\ 1 \\ 0 \\ 0 \end{pmatrix}.$$

(b) $\Phi = \begin{pmatrix} -6 & 2e^t & 3e^{3t} & -e^{4t} \\ -4 & e^t & e^{3t} & e^{4t} \\ 1 & 0 & 2e^{3t} & 0 \\ 2 & 0 & e^{3t} & 0 \end{pmatrix}$, $\quad \Phi^{-1} = \begin{pmatrix} 0 & 0 & -\frac{1}{3} & \frac{2}{3} \\ \frac{1}{3}e^{-t} & \frac{1}{3}e^{-t} & -2e^{-t} & \frac{8}{3}e^{-t} \\ 0 & 0 & \frac{2}{3}e^{-3t} & -\frac{1}{3}e^{-3t} \\ -\frac{1}{3}e^{-4t} & \frac{2}{3}e^{-4t} & 0 & \frac{1}{3}e^{-4t} \end{pmatrix}$

(c) $\Phi^{-1}(t)\mathbf{F}(t) = \begin{pmatrix} \frac{2}{3} - \frac{1}{3}e^{2t} \\ \frac{1}{3}e^{-2t} + \frac{8}{3}e^{-t} - 2e^t + \frac{1}{3}t \\ -\frac{1}{3}e^{-3t} + \frac{2}{3}e^{-t} \\ \frac{2}{3}e^{-5t} + \frac{1}{3}e^{-4t} - \frac{1}{3}te^{-3t} \end{pmatrix}$,

$$\int \Phi^{-1}(t)\mathbf{F}(t)\,dt = \begin{pmatrix} -\frac{1}{6}e^{2t} + \frac{2}{3}t \\ -\frac{1}{6}e^{-2t} - \frac{8}{3}e^{-t} - 2e^t + \frac{1}{6}t^2 \\ \frac{1}{9}e^{-3t} - \frac{2}{3}e^{-t} \\ -\frac{2}{15}e^{-5t} - \frac{1}{12}e^{-4t} + \frac{1}{27}e^{-3t} + \frac{1}{9}te^{-3t} \end{pmatrix},$$

$$\Phi(t)\int \Phi^{-1}(t)\mathbf{F}(t)\,dt = \begin{pmatrix} -5e^{2t} - \frac{1}{5}e^{-t} - \frac{1}{27}e^t - \frac{1}{9}te^t + \frac{1}{3}t^2e^t - 4t - \frac{59}{12} \\ -2e^{2t} - \frac{3}{10}e^{-t} + \frac{1}{27}e^t + \frac{1}{9}te^t + \frac{1}{6}t^2e^t - \frac{8}{3}t - \frac{95}{36} \\ -\frac{3}{2}e^{2t} + \frac{2}{3}t + \frac{2}{9} \\ -e^{2t} + \frac{4}{3}t - \frac{1}{9} \end{pmatrix},$$

$$\Phi(t)\mathbf{C} = \begin{pmatrix} -6c_1 + 2c_2e^t + 3c_3e^{3t} - c_4e^{4t} \\ -4c_1 + c_2e^t + c_3e^{3t} + c_4e^{4t} \\ c_1 + 2c_3e^{3t} \\ 2c_1 + c_3e^{3t} \end{pmatrix},$$

$$\Phi(t)\mathbf{C} + \Phi(t)\int \Phi^{-1}(t)\mathbf{F}(t)\,dt$$

$$= \begin{pmatrix} -6c_1 + 2c_2e^t + 3c_3e^{3t} - c_4e^{4t} \\ -4c_1 + c_2e^t + c_3e^{3t} + c_4e^{4t} \\ c_1 + 2c_3e^{3t} \\ 2c_1 + c_3e^{3t} \end{pmatrix} + \begin{pmatrix} -5e^{2t} - \frac{1}{5}e^{-t} - \frac{1}{27}e^t - \frac{1}{9}te^t + \frac{1}{3}t^2e^t - 4t - \frac{59}{12} \\ -2e^{2t} - \frac{3}{10}e^{-t} + \frac{1}{27}e^t + \frac{1}{9}te^t + \frac{1}{6}t^2e^t - \frac{8}{3}t - \frac{95}{36} \\ -\frac{3}{2}e^{2t} + \frac{2}{3}t + \frac{2}{9} \\ -e^{2t} + \frac{4}{3}t - \frac{1}{9} \end{pmatrix}$$

(d) $\mathbf{X}(t) = c_1 \begin{pmatrix} -6 \\ -4 \\ 1 \\ 2 \end{pmatrix} + c_2 \begin{pmatrix} 2 \\ 1 \\ 0 \\ 0 \end{pmatrix} + c_3 \begin{pmatrix} 3 \\ 1 \\ 2 \\ 1 \end{pmatrix} + c_4 \begin{pmatrix} -1 \\ 1 \\ 0 \\ 0 \end{pmatrix}$

$+ \begin{pmatrix} -5e^{2t} - \frac{1}{5}e^{-t} - \frac{1}{27}e^t - \frac{1}{9}te^t + \frac{1}{3}t^2e^t - 4t - \frac{59}{12} \\ -2e^{2t} - \frac{3}{10}e^{-t} + \frac{1}{27}e^t + \frac{1}{9}te^t + \frac{1}{6}t^2e^t - \frac{8}{3}t - \frac{95}{36} \\ -\frac{3}{2}e^{2t} + \frac{2}{3}t + \frac{2}{9} \\ -e^{2t} + \frac{4}{3}t - \frac{1}{9} \end{pmatrix}$

Exercises 8.4

3. For

$$\mathbf{A} = \begin{pmatrix} 1 & 1 & 1 \\ 1 & 1 & 1 \\ -2 & -2 & -2 \end{pmatrix}$$

we have

$$\mathbf{A}^2 = \begin{pmatrix} 1 & 1 & 1 \\ 1 & 1 & 1 \\ -2 & -2 & -2 \end{pmatrix} \begin{pmatrix} 1 & 1 & 1 \\ 1 & 1 & 1 \\ -2 & -2 & -2 \end{pmatrix} = \begin{pmatrix} 0 & 0 & 0 \\ 0 & 0 & 0 \\ 0 & 0 & 0 \end{pmatrix}.$$

Thus, $\mathbf{A}^3 = \mathbf{A}^4 = \mathbf{A}^5 = \cdots = \mathbf{0}$ and

$$e^{\mathbf{A}t} = \mathbf{I} + \mathbf{A}t = \begin{pmatrix} 1 & 0 & 0 \\ 0 & 1 & 0 \\ 0 & 0 & 1 \end{pmatrix} + \begin{pmatrix} t & t & t \\ t & t & t \\ -2t & -2t & -2t \end{pmatrix} = \begin{pmatrix} t+1 & t & t \\ t & t+1 & t \\ -2t & -2t & -2t+1 \end{pmatrix}.$$

6. For $\mathbf{A} = \begin{pmatrix} 0 & 1 \\ 1 & 0 \end{pmatrix}$ we have

$$e^{t\mathbf{A}} = \begin{pmatrix} \cosh t & \sinh t \\ \sinh t & \cosh t \end{pmatrix}.$$

Then

$$\mathbf{X} = \begin{pmatrix} \cosh t & \sinh t \\ \sinh t & \cosh t \end{pmatrix} \begin{pmatrix} c_1 \\ c_2 \end{pmatrix} = c_1 \begin{pmatrix} \cosh t \\ \sinh t \end{pmatrix} + c_2 \begin{pmatrix} \sinh t \\ \cosh t \end{pmatrix}.$$

9. To solve

$$\mathbf{X}' = \begin{pmatrix} 1 & 0 \\ 0 & 2 \end{pmatrix} \mathbf{X} + \begin{pmatrix} 3 \\ -1 \end{pmatrix}$$

we identify $t_0 = 0$, $\mathbf{F}(s) = \begin{pmatrix} 3 \\ -1 \end{pmatrix}$, and use the results of Problem 1 and equation (5) in the text.

$$\mathbf{X}(t) = e^{\mathbf{A}t}\mathbf{C} + e^{\mathbf{A}t}\int_{t_0}^{t} e^{-\mathbf{A}s}\mathbf{F}(s)\,ds$$

$$= \begin{pmatrix} e^t & 0 \\ 0 & e^{2t} \end{pmatrix}\begin{pmatrix} c_1 \\ c_2 \end{pmatrix} + \begin{pmatrix} e^t & 0 \\ 0 & e^{2t} \end{pmatrix}\int_0^t \begin{pmatrix} e^{-s} & 0 \\ 0 & e^{-2s} \end{pmatrix}\begin{pmatrix} 3 \\ -1 \end{pmatrix}\,ds$$

$$= \begin{pmatrix} c_1 e^t \\ c_2 e^{2t} \end{pmatrix} + \begin{pmatrix} e^t & 0 \\ 0 & e^{2t} \end{pmatrix}\int_0^t \begin{pmatrix} 3e^{-s} \\ -e^{-2s} \end{pmatrix}\,ds$$

$$= \begin{pmatrix} c_1 e^t \\ c_2 e^{2t} \end{pmatrix} + \begin{pmatrix} e^t & 0 \\ 0 & e^{2t} \end{pmatrix}\begin{pmatrix} -3e^{-s} \\ \frac{1}{2}e^{-2s} \end{pmatrix}\Big|_0^t$$

$$= \begin{pmatrix} c_1 e^t \\ c_2 e^{2t} \end{pmatrix} + \begin{pmatrix} e^t & 0 \\ 0 & c^{2t} \end{pmatrix}\begin{pmatrix} -3e^{-t} - 3 \\ \frac{1}{2}c^{-2t} & \frac{1}{2} \end{pmatrix}$$

$$= \begin{pmatrix} c_1 e^t \\ c_2 e^{2t} \end{pmatrix} + \begin{pmatrix} -3 - 3e^t \\ \frac{1}{2} - \frac{1}{2}e^{2t} \end{pmatrix} = c_3 \begin{pmatrix} 1 \\ 0 \end{pmatrix}e^t + c_4 \begin{pmatrix} 0 \\ 1 \end{pmatrix}e^{2t} + \begin{pmatrix} -3 \\ \frac{1}{2} \end{pmatrix}.$$

12. To solve

$$\mathbf{X}' = \begin{pmatrix} 0 & 1 \\ 1 & 0 \end{pmatrix}\mathbf{X} + \begin{pmatrix} \cosh t \\ \sinh t \end{pmatrix}$$

we identify $t_0 = 0$, $\mathbf{F}(s) = \begin{pmatrix} \cosh t \\ \sinh t \end{pmatrix}$, and use equation (5) in the text and the value of $e^{t\mathbf{A}}$ given in the solution of Problem 6.

$$\mathbf{X}(t) = e^{\mathbf{A}t}\mathbf{C} + e^{\mathbf{A}t}\int_{t_0}^{t} e^{-\mathbf{A}s}\mathbf{F}(s)\,ds$$

$$= \begin{pmatrix} \cosh t & \sinh t \\ \sinh t & \cosh t \end{pmatrix}\begin{pmatrix} c_1 \\ c_2 \end{pmatrix} + \begin{pmatrix} \cosh t & \sinh t \\ \sinh t & \cosh t \end{pmatrix}\int_0^t \begin{pmatrix} \cosh s & -\sinh s \\ -\sinh s & \cosh s \end{pmatrix}\begin{pmatrix} \cosh s \\ \sinh s \end{pmatrix}\,ds$$

$$= \begin{pmatrix} c_1 \cosh t + c_2 \sinh t \\ c_1 \sinh t + c_2 \cosh t \end{pmatrix} + \begin{pmatrix} \cosh t & \sinh t \\ \sinh t & \cosh t \end{pmatrix}\int_0^t \begin{pmatrix} 1 \\ 0 \end{pmatrix}\,ds$$

$$= \begin{pmatrix} c_1 \cosh t + c_2 \sinh t \\ c_1 \sinh t + c_2 \cosh t \end{pmatrix} + \begin{pmatrix} \cosh t & \sinh t \\ \sinh t & \cosh t \end{pmatrix}\begin{pmatrix} s \\ 0 \end{pmatrix}\Big|_0^t$$

$$= \begin{pmatrix} c_1 \cosh t + c_2 \sinh t \\ c_1 \sinh t + c_2 \cosh t \end{pmatrix} + \begin{pmatrix} \cosh t & \sinh t \\ \sinh t & \cosh t \end{pmatrix}\begin{pmatrix} t \\ 0 \end{pmatrix}$$

$$= \begin{pmatrix} c_1 \cosh t + c_2 \sinh t \\ c_1 \sinh t + c_2 \cosh t \end{pmatrix} + \begin{pmatrix} t \cosh t \\ t \sinh t \end{pmatrix} = c_1 \begin{pmatrix} \cosh t \\ \sinh t \end{pmatrix} + c_2 \begin{pmatrix} \sinh t \\ \cosh t \end{pmatrix} + t \begin{pmatrix} \cosh t \\ \sinh t \end{pmatrix}.$$

15. Solving

$$\begin{vmatrix} 2-\lambda & 1 \\ -3 & 6-\lambda \end{vmatrix} = \lambda^2 - 8\lambda + 15 = (\lambda - 3)(\lambda - 5) = 0$$

we find eigenvalues $\lambda_1 = 3$ and $\lambda_2 = 5$. Corresponding eigenvectors are

$$\mathbf{K}_1 = \begin{pmatrix} 1 \\ 1 \end{pmatrix} \quad \text{and} \quad \mathbf{K}_2 = \begin{pmatrix} 1 \\ 3 \end{pmatrix}.$$

Then $\qquad \mathbf{P} = \begin{pmatrix} 1 & 1 \\ 1 & 3 \end{pmatrix}, \quad \mathbf{P}^{-1} = \begin{pmatrix} 3/2 & -1/2 \\ -1/2 & 1/2 \end{pmatrix}, \quad \text{and} \quad \mathbf{D} = \begin{pmatrix} 3 & 0 \\ 0 & 5 \end{pmatrix},$

so $\qquad\qquad\qquad\qquad\qquad \mathbf{P}\mathbf{D}\mathbf{P}^{-1} = \begin{pmatrix} 2 & 1 \\ -3 & 6 \end{pmatrix}.$

18. From equation (3) in the text

$$e^{\mathbf{D}t} = \begin{pmatrix} 1 & 0 & \cdots & 0 \\ 0 & 1 & \cdots & 0 \\ \vdots & \vdots & \ddots & \vdots \\ 0 & 0 & \cdots & 1 \end{pmatrix} + \begin{pmatrix} \lambda_1 & 0 & \cdots & 0 \\ 0 & \lambda_2 & \cdots & 0 \\ \vdots & \vdots & \ddots & \vdots \\ 0 & 0 & \cdots & \lambda_n \end{pmatrix} + \frac{1}{2!}t^2 \begin{pmatrix} \lambda_1^2 & 0 & \cdots & 0 \\ 0 & \lambda_2^2 & \cdots & 0 \\ \vdots & \vdots & \ddots & \vdots \\ 0 & 0 & \cdots & \lambda_n^2 \end{pmatrix}$$

$$+ \frac{1}{3!}t^3 \begin{pmatrix} \lambda_1^3 & 0 & \cdots & 0 \\ 0 & \lambda_2^3 & \cdots & 0 \\ \vdots & \vdots & \ddots & \vdots \\ 0 & 0 & \cdots & \lambda_n^3 \end{pmatrix} + \cdots$$

$$= \begin{pmatrix} 1 + \lambda_1 t + \frac{1}{2!}(\lambda_1 t)^2 + \cdots & 0 & \cdots & 0 \\ 0 & 1 + \lambda_2 t + \frac{1}{2!}(\lambda_2 t)^2 + \cdots & \cdots & 0 \\ \vdots & \vdots & \ddots & \vdots \\ 0 & 0 & \cdots & 1 + \lambda_n t + \frac{1}{2!}(\lambda_n t)^2 + \cdots \end{pmatrix}$$

$$= \begin{pmatrix} e^{\lambda_1 t} & 0 & \cdots & 0 \\ 0 & e^{\lambda_2 t} & \cdots & 0 \\ \vdots & \vdots & \ddots & \vdots \\ 0 & 0 & \cdots & e^{\lambda_n t} \end{pmatrix}$$

115

Chapter 8 Review Exercises

3. We have $\det(\mathbf{A} - \lambda\mathbf{I}) = (\lambda - 1)^2 = 0$ and $\mathbf{K} = \begin{pmatrix} 1 \\ -1 \end{pmatrix}$. A solution to $(\mathbf{A} - \lambda\mathbf{I})\mathbf{P} = \mathbf{K}$ is $\mathbf{P} = \begin{pmatrix} 0 \\ 1 \end{pmatrix}$

so that
$$\mathbf{X} = c_1 \begin{pmatrix} 1 \\ -1 \end{pmatrix} e^t + c_2 \left[\begin{pmatrix} 1 \\ -1 \end{pmatrix} te^t + \begin{pmatrix} 0 \\ 1 \end{pmatrix} e^t \right].$$

6. We have $\det(\mathbf{A} - \lambda\mathbf{I}) = \lambda^2 - 2\lambda + 2 = 0$. For $\lambda = 1 + i$ we obtain $\mathbf{K}_1 = \begin{pmatrix} 3 - i \\ 2 \end{pmatrix}$ and

$$\mathbf{X}_1 = \begin{pmatrix} 3 - i \\ 2 \end{pmatrix} e^{(1+i)t} = \begin{pmatrix} 3\cos t + \sin t \\ 2\cos t \end{pmatrix} e^t + i \begin{pmatrix} -\cos t + 3\sin t \\ 2\sin t \end{pmatrix} e^t.$$

Then
$$\mathbf{X} = c_1 \begin{pmatrix} 3\cos t + \sin t \\ 2\cos t \end{pmatrix} e^t + c_2 \begin{pmatrix} -\cos t + 3\sin t \\ 2\sin t \end{pmatrix} e^t.$$

9. We have
$$\mathbf{X}_c = c_1 \begin{pmatrix} 1 \\ 0 \end{pmatrix} e^{2t} + c_2 \begin{pmatrix} 4 \\ 1 \end{pmatrix} e^{4t}.$$

Then
$$\mathbf{\Phi} = \begin{pmatrix} e^{2t} & 4e^{4t} \\ 0 & e^{4t} \end{pmatrix}, \quad \mathbf{\Phi}^{-1} = \begin{pmatrix} e^{-2t} & -4e^{-2t} \\ 0 & e^{-4t} \end{pmatrix},$$

and
$$\mathbf{U} = \int \mathbf{\Phi}^{-1} \mathbf{F}\, dt = \int \begin{pmatrix} 2e^{-2t} - 64te^{-2t} \\ 16te^{-4t} \end{pmatrix} dt = \begin{pmatrix} 15e^{-2t} + 32te^{-2t} \\ -e^{-4t} - 4te^{-4t} \end{pmatrix},$$

so that
$$\mathbf{X}_p = \mathbf{\Phi}\mathbf{U} = \begin{pmatrix} 11 + 16t \\ -1 - 4t \end{pmatrix}.$$

12. We have
$$\mathbf{X}_c = c_1 \begin{pmatrix} 1 \\ -1 \end{pmatrix} e^{2t} + c_2 \left[\begin{pmatrix} 1 \\ -1 \end{pmatrix} te^{2t} + \begin{pmatrix} 1 \\ 0 \end{pmatrix} e^{2t} \right].$$

Then
$$\mathbf{\Phi} = \begin{pmatrix} e^{2t} & te^{2t} + e^{2t} \\ -e^{2t} & -te^{2t} \end{pmatrix}, \quad \mathbf{\Phi}^{-1} = \begin{pmatrix} -te^{-2t} & -te^{-2t} - e^{-2t} \\ e^{-2t} & e^{-2t} \end{pmatrix},$$

and
$$\mathbf{U} = \int \mathbf{\Phi}^{-1} \mathbf{F}\, dt = \int \begin{pmatrix} t - 1 \\ -1 \end{pmatrix} dt = \begin{pmatrix} \frac{1}{2}t^2 - t \\ -t \end{pmatrix},$$

so that
$$\mathbf{X}_p = \mathbf{\Phi}\mathbf{U} = \begin{pmatrix} -1/2 \\ 1/2 \end{pmatrix} t^2 e^{2t} + \begin{pmatrix} -2 \\ 1 \end{pmatrix} te^{2t}.$$

9 Numerical Methods for Ordinary Differential Equations

——————— Exercises 9.1 ———————

3.

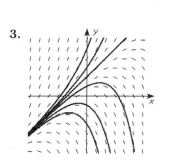

6. Setting $x + y = c$ we obtain the isoclines $y = -x + c$.

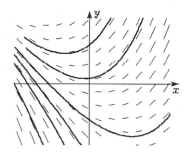

9. Setting $0.2x^2 + y = c$ we obtain the isoclines $y = c - 0.2x^2$.

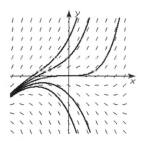

12. Setting $1 - y/x = c$ we obtain the isoclines $y = (1 - c)x$.

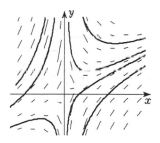

Exercises 9.2

All tables in this chapter were constructed in a spreadsheet program which does not support subscripts. Consequently, x_n and y_n will be indicated as $x(n)$ and $y(n)$, respectively.

3.

h = 0.1			h = 0.05	
$x(n)$	$y(n)$		$x(n)$	$y(n)$
1.00	5.0000		1.00	5.0000
1.10	3.8000		1.05	4.4000
1.20	2.9800		1.10	3.8950
1.30	2.4260		1.15	3.4708
1.40	2.0582		1.20	3.1151
1.50	1.8207		1.25	2.8179
			1.30	2.5702
			1.35	2.3647
			1.40	2.1950
			1.45	2.0557
			1.50	1.9424

6.

h = 0.1			h = 0.05	
$x(n)$	$y(n)$		$x(n)$	$y(n)$
0.00	1.0000		0.00	1.0000
0.10	1.1000		0.05	1.0500
0.20	1.2220		0.10	1.1053
0.30	1.3753		0.15	1.1668
0.40	1.5735		0.20	1.2360
0.50	1.8371		0.25	1.3144
			0.30	1.4039
			0.35	1.5070
			0.40	1.6267
			0.45	1.7670
			0.50	1.9332

9.

h = 0.1			h = 0.05	
$x(n)$	$y(n)$		$x(n)$	$y(n)$
0.00	0.5000		0.00	0.5000
0.10	0.5250		0.05	0.5125
0.20	0.5431		0.10	0.5232
0.30	0.5548		0.15	0.5322
0.40	0.5613		0.20	0.5395
0.50	0.5639		0.25	0.5452
			0.30	0.5496
			0.35	0.5527
			0.40	0.5547
			0.45	0.5559
			0.50	0.5565

12.

h =	0.1
x(n)	y(n}
0.00	0.5000
0.10	0.5250
0.20	0.5499
0.30	0.5747
0.40	0.5991
0.50	0.6231

h =	0.05
x(n)	y(n}
0.00	0.5000
0.05	0.5125
0.10	0.5250
0.15	0.5375
0.20	0.5499
0.25	0.5623
0.30	0.5746
0.35	0.5868
0.40	0.5989
0.45	0.6109
0.50	0.6228

15. (a)

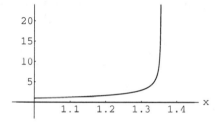

(b)

h=0.1	EULER	IMPROVED EULER
x(n)	y(n)	y(n)
1.00	1.0000	1.0000
1.10	1.2000	1.2469
1.20	1.4938	1.6668
1.30	1.9711	2.6427
1.40	2.9060	8.7988

18. (a) Using the improved Euler method we obtain $y(0.1) \approx y_1 = 1.22$.

(b) Using $y''' = 8e^{2x}$ we see that the local truncation error is

$$y'''(c) \frac{h^3}{6} = 8e^{2c} \frac{(0.1)^3}{6} = 0.001333e^{2c}.$$

Since e^{2x} is an increasing function, $e^{2c} \le e^{2(0.1)} = e^{0.2}$ for $0 \le c \le 0.1$. Thus an upper bound for the local truncation error is $0.001333e^{0.2} = 0.001628$.

(c) Since $y(0.1) = e^{0.2} = 1.221403$, the actual error is $y(0.1) - y_1 = 0.001403$ which is less than 0.001628.

(d) Using the improved Euler method with $h = 0.05$ we obtain $y(0.1) \approx y_2 = 1.221025$.

119

(e) The error in (d) is $1.221403 - 1.221025 = 0.000378$. With global truncation error $O(h^2)$, when the step size is halved we expect the error for $h = 0.05$ to be one-fourth the error for $h = 0.1$. Comparing 0.000378 with 0.001403 we see that this is the case.

21. (a) Using $y'' = 38e^{-3(x-1)}$ we see that the local truncation error is

$$y''(c)\frac{h^2}{2} = 38e^{-3(c-1)}\frac{h^2}{2} = 19h^2 e^{-3(c-1)}.$$

(b) Since $e^{-3(x-1)}$ is a decreasing function for $1 \le x \le 1.5$, $e^{-3(c-1)} \le e^{-3(1-1)} = 1$ for $1 \le c \le 1.5$ and

$$y''(c)\frac{h^2}{2} \le 19(0.1)^2(1) = 0.19.$$

(c) Using the Euler method with $h = 0.1$ we obtain $y(1.5) \approx 1.8207$. With $h = 0.05$ we obtain $y(1.5) \approx 1.9424$.

(d) Since $y(1.5) = 2.0532$, the error for $h = 0.1$ is $E_{0.1} = 0.2325$, while the error for $h = 0.05$ is $E_{0.05} = 0.1109$. With global truncation error $O(h)$ we expect $E_{0.1}/E_{0.05} \approx 2$. We actually have $E_{0.1}/E_{0.05} = 2.10$.

24. (a) Using $y''' = \dfrac{2}{(x+1)^3}$ we see that the local truncation error is

$$y'''(c)\frac{h^3}{6} = \frac{1}{(c+1)^3}\frac{h^3}{3}.$$

(b) Since $\dfrac{1}{(x+1)^3}$ is a decreasing function for $0 \le x \le 0.5$, $\dfrac{1}{(c+1)^3} \le \dfrac{1}{(0+1)^3} = 1$ for $0 \le c \le 0.5$ and

$$y'''(c)\frac{h^3}{6} \le (1)\frac{(0.1)^3}{3} = 0.000333.$$

(c) Using the improved Euler method with $h = 0.1$ we obtain $y(0.5) \approx 0.405281$. With $h = 0.05$ we obtain $y(0.5) \approx 0.405419$.

(d) Since $y(0.5) = 0.405465$, the error for $h = 0.1$ is $E_{0.1} = 0.000184$, while the error for $h = 0.05$ is $E_{0.05} = 0.000046$. With global truncation error $O(h^2)$ we expect $E_{0.1}/E_{0.05} \approx 4$. We actually have $E_{0.1}/E_{0.05} = 3.98$.

_____ **Exercises 9.3** _____

3.

x(n)	y(n)
1.00	5.0000
1.10	3.9724
1.20	3.2284
1.30	2.6945
1.40	2.3163
1.50	2.0533

6.

x(n)	y(n)
0.00	1.0000
0.10	1.1115
0.20	1.2530
0.30	1.4397
0.40	1.6961
0.50	2.0670

9.

x(n)	y(n)
0.00	0.5000
0.10	0.5213
0.20	0.5358
0.30	0.5443
0.40	0.5482
0.50	0.5493

12.

x(n)	y(n)
0.00	0.5000
0.10	0.5250
0.20	0.5498
0.30	0.5744
0.40	0.5987
0.50	0.6225

15. (a)

x(n)	h = 0.05 y(n)	h = 0.1 y(n)
1.00	1.0000	1.0000
1.05	1.1112	
1.10	1.2511	1.2511
1.15	1.4348	
1.20	1.6934	1.6934
1.25	2.1047	
1.30	2.9560	2.9425
1.35	7.8981	
1.40	1.06E+15	903.0282

(b)

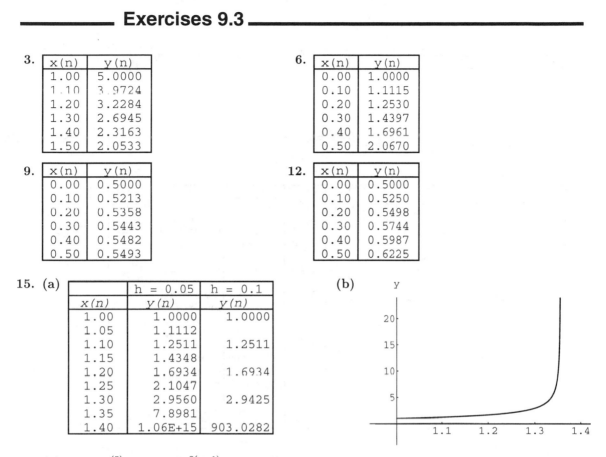

18. (a) Using $y^{(5)} = -1026e^{-3(x-1)}$ we see that the local truncation error is

$$\left| y^{(5)}(c)\frac{h^5}{120} \right| = 8.55h^5 e^{-3(c-1)}.$$

(b) Since $e^{-3(x-1)}$ is a decreasing function for $1 \le x \le 1.5$, $e^{-3(c-1)} \le e^{-3(1-1)} = 1$ for $1 \le c \le 1.5$ and

$$y^{(5)}(c)\frac{h^5}{120} \le 8.55(0.1)^5(1) = 0.0000855.$$

(c) Using the fourth-order Runge-Kutta method with $h = 0.1$ we obtain $y(1.5) \approx 2.053338827$. With $h = 0.05$ we obtain $y(1.5) \approx 2.053222989$.

Exercises 9.4

3.

$x(n)$	$y(n)$	
0.00	1.0000	initial condition
0.20	0.7328	Runge-Kutta
0.40	0.6461	Runge-Kutta
0.60	0.6585	Runge-Kutta
	0.7332	predictor
0.80	0.7232	corrector

6.

$x(n)$	$y(n)$	
0.00	1.0000	initial condition
0.20	1.4414	Runge-Kutta
0.40	1.9719	Runge-Kutta
0.60	2.6028	Runge-Kutta
	3.3483	predictor
0.80	3.3486	corrector
	4.2276	predictor
1.00	4.2280	corrector

$x(n)$	$y(n)$	
0.00	1.0000	initial condition
0.10	1.2102	Runge-Kutta
0.20	1.4414	Runge-Kutta
0.30	1.6949	Runge-Kutta
	1.9719	predictor
0.40	1.9719	corrector
	2.2740	predictor
0.50	2.2740	corrector
	2.6028	predictor
0.60	2.6028	corrector
	2.9603	predictor
0.70	2.9603	corrector
	3.3486	predictor
0.80	3.3486	corrector
	3.7703	predictor
0.90	3.7703	corrector
	4.2280	predictor
1.00	4.2280	corrector

Exercises 9.5

3. The substitution $y' = u$ leads to the system

$$y' = u, \qquad u' = 4u - 4y.$$

Using formulas (5) and (6) in the text with x corresponding to t, y corresponding to x, and u corresponding to y, we obtain

122

								Runge-Kutta method with h=0.2		
m1	m2	m3	m4	k1	k2	k3	k4	x	y	u
								0.00	-2.0000	1.0000
0.2000	0.4400	0.5280	0.9072	2.4000	3.2800	3.5360	4.8064	0.20	-1.4928	4.4731

								Runge-Kutta method with h=0.1		
m1	m2	m3	m4	k1	k2	k3	k4	x	y	u
								0.00	-2.0000	1.0000
0.1000	0.1600	0.1710	0.2452	1.2000	1.4200	1.4520	1.7124	0.10	-1.8321	2.4427
0.2443	0.3298	0.3444	0.4487	1.7099	2.0031	2.0446	2.3900	0.20	-1.4919	4.4753

6.

								Runge-Kutta method with h=0.1		
m1	m2	m3	m4	k1	k2	k3	k4	t	i1	i2
								0.00	0.0000	0.0000
10.0000	0.0000	12.5000	-20.0000	0.0000	5.0000	-5.0000	22.5000	0.10	2.5000	3.7500
8.7500	-2.5000	13.4375	-28.7500	-5.0000	4.3750	-10.6250	29.6875	0.20	2.8125	5.7813
10.1563	-4.3750	17.0703	-40.0000	-8.7500	5.0781	-16.0156	40.3516	0.30	2.0703	7.4023
13.2617	-6.3672	22.9443	-55.1758	-12.7344	6.6309	-22.5488	55.3076	0.40	0.6104	9.1919
17.9712	-8.8867	31.3507	-75.9326	-17.7734	8.9856	-31.2024	75.9821	0.50	-1.5619	11.4877

9.

								Runge-Kutta method with h=0.2		
m1	m2	m3	m4	k1	k2	k3	k4	t	x	y
								0.00	-3.0000	5.0000
-1.0000	-0.9200	-0.9080	-0.8176	0.6000	-0.7200	-0.7120	-0.8216	0.20	-3.9123	4.2857

								Runge-Kutta method with h=0.1		
m1	m2	m3	m4	k1	k2	k3	k4	t	x	y
								0.00	-3.0000	5.0000
-0.5000	0.4800	-0.4785	-0.4571	-0.3000	-0.3300	-0.3290	-0.3579	0.10	-3.4790	4.6707
-0.4571	-0.4342	-0.4328	-0.4086	-0.3579	-0.3858	-0.3846	-0.4112	0.20	-3.9123	4.2857

12. Solving for x' and y' we obtain the system

$$x' = \frac{1}{2}y - 3t^2 + 2t - 5$$

$$y' = -\frac{1}{2}y + 3t^2 + 2t + 5.$$

m1	m2	m3	m4	k1	k2	k3	k4	t	x	y
							Runge-Kutta method with h=0.2			
								0.00	3.0000	-1.0000
-1.1000	-1.0110	-1.0115	-0.9349	1.1000	1.0910	1.0915	1.0949	0.20	1.9867	0.0933

m1	m2	m3	m4	k1	k2	k3	k4	t	x	y
							Runge-Kutta method with h=0.1			
								0.00	3.0000	-1.0000
-0.5500	-0.5270	-0.5271	-0.5056	0.5500	0.5470	0.5471	0.5456	0.10	2.4727	-0.4527
-0.5056	-0.4857	-0.4857	-0.4673	0.5456	0.5457	0.5457	0.5473	0.20	1.9867	0.0933

Exercises 9.6

3. We identify $P(x) = 2$, $Q(x) = 1$, $f(x) = 5x$, and $h = (1-0)/5 = 0.2$. Then the finite difference equation is

$$1.2y_{i+1} - 1.96y_i + 0.8y_{i-1} = 0.04(5x_i).$$

The solution of the corresponding linear system gives

x	0.0	0.2	0.4	0.6	0.8	1.0
y	0.0000	-0.2259	-0.3356	-0.3308	-0.2167	0.0000

6. We identify $P(x) = 5$, $Q(x) = 0$, $f(x) = 4\sqrt{x}$, and $h = (2-1)/6 = 0.1667$. Then the finite difference equation is

$$1.4167y_{i+1} - 2y_i + 0.5833y_{i-1} = 0.2778(4\sqrt{x_i}).$$

The solution of the corresponding linear system gives

x	1.0000	1.1667	1.3333	1.5000	1.6667	1.8333	2.0000
y	1.0000	-0.5918	-1.1626	-1.3070	-1.2704	-1.1541	-1.0000

9. We identify $P(x) = 1 - x$, $Q(x) = x$, $f(x) = x$, and $h = (1-0)/10 = 0.1$. Then the finite difference equation is

$$[1 + 0.05(1 - x_i)]y_{i+1} + [-2 + 0.01x_i]y_i + [1 - 0.05(1 - x_i)]y_{i-1} = 0.01x_i.$$

The solution of the corresponding linear system gives

x	0.0	0.1	0.2	0.3	0.4	0.5	0.6
y	0.0000	0.2660	0.5097	0.7357	0.9471	1.1465	1.3353

0.7	0.8	0.9	1.0
1.5149	1.6855	1.8474	2.0000

12. We identify $P(r) = 2/r$, $Q(r) = 0$, $f(r) = 0$, and $h = (4-1)/6 = 0.5$. Then the finite difference equation is

$$\left(1 + \frac{0.5}{r_i}\right)u_{i+1} - 2u_i + \left(1 - \frac{0.5}{r_i}\right)u_{i-1} = 0.$$

124

The solution of the corresponding linear system gives

r	1.0	1.5	2.0	2.5	3.0	3.5	4.0
u	50.0000	72.2222	83.3333	90.0000	94.4444	97.6190	100.0000

Chapter 9 Review Exercises

3.

$h=0.1$		IMPROVED	RUNGE	$h=0.05$		IMPROVED	RUNGE
$x(n)$	EULER	EULER	KUTTA	$x(n)$	EULER	EULER	KUTTA
1.00	2.0000	2.0000	2.0000	1.00	2.0000	2.0000	2.0000
1.10	2.1386	2.1549	2.1556	1.05	2.0693	2.0735	2.0736
1.20	2.3097	2.3439	2.3454	1.10	2.1469	2.1554	2.1556
1.30	2.5136	2.5672	2.5695	1.15	2.2328	2.2459	2.2462
1.40	2.7504	2.8246	2.8278	1.20	2.3272	2.3450	2.3454
1.50	3.0201	3.1157	3.1197	1.25	2.4299	2.4527	2.4532
				1.30	2.5409	2.5689	2.5695
				1.35	2.6604	2.6937	2.6944
				1.40	2.7883	2.8269	2.8278
				1.45	2.9245	2.9686	2.9696
				1.50	3.0690	3.1187	3.1197

6.

$h=0.1$		IMPROVED	RUNGE	$h=0.05$		IMPROVED	RUNGE
$x(n)$	EULER	EULER	KUTTA	$x(n)$	EULER	EULER	KUTTA
1.00	1.0000	1.0000	1.0000	1.00	1.0000	1.0000	1.0000
1.10	1.2000	1.2380	1.2415	1.05	1.1000	1.1091	1.1095
1.20	1.4760	1.5910	1.6036	1.10	1.2183	1.2405	1.2415
1.30	1.8710	2.1524	2.1909	1.15	1.3595	1.4010	1.4029
1.40	2.4643	3.1458	3.2745	1.20	1.5300	1.6001	1.6036
1.50	3.4165	5.2510	5.8338	1.25	1.7389	1.8523	1.8586
				1.30	1.9988	2.1799	2.1911
				1.35	2.3284	2.6197	2.6401
				1.40	2.7567	3.2360	3.2755
				1.45	3.3296	4.1528	4.2363
				1.50	4.1253	5.6404	5.8446

9. Using $x_0 = 1$, $y_0 = 2$, and $h = 0.1$ we have

$$x_1 = x_0 + h(x_0 + y_0) = 1 + 0.1(1 + 2) = 1.3$$

$$y_1 = y_0 + h(x_0 - y_0) = 2 + 0.1(1 - 2) = 1.9$$

and

$$x_2 = x_1 + h(x_1 + y_1) = 1.3 + 0.1(1.3 + 1.9) = 1.62$$

$$y_2 = y_1 + h(x_1 - y_1) = 1.9 + 0.1(1.3 - 1.9) = 1.84.$$

Thus, $x(0.2) \approx 1.62$ and $y(0.2) \approx 1.84$.

Appendix

────────── **Appendix I** ──────────

3. If $t = x^3$, then $dt = 3x^2 \, dx$ and $x^4 \, dx = \frac{1}{3}t^{2/3} \, dt$. Now

$$\int_0^\infty x^4 e^{-x^3} \, dx = \int_0^\infty \frac{1}{3}t^{2/3} e^{-t} \, dt = \frac{1}{3}\int_0^\infty t^{2/3} e^{-t} \, dt$$

$$= \frac{1}{3}\Gamma\left(\frac{5}{3}\right) = \frac{1}{3}(0.89) \approx 0.297.$$

6. For $x > 0$

$$\Gamma(x+1) = \int_0^\infty t^x e^{-t} dt$$

$u = t^x$	$dv = e^{-t}\,dt$
$du = xt^{x-1}\,dt$	$v = -e^{-t}$

$$= -t^x e^{-t}\,\Big|_0^\infty - \int_0^\infty xt^{x-1}(-e^{-t})\,dt$$

$$= x\int_0^\infty t^{x-1}e^{-t}dt = x\Gamma(x).$$

────────── **Appendix II** ──────────

3. (a) $\mathbf{AB} = \begin{pmatrix} -2-9 & 12-6 \\ 5+12 & -30+8 \end{pmatrix} = \begin{pmatrix} -11 & 6 \\ 17 & -22 \end{pmatrix}$

(b) $\mathbf{BA} = \begin{pmatrix} -2-30 & 3+24 \\ 6-10 & -9+8 \end{pmatrix} = \begin{pmatrix} -32 & 27 \\ -4 & -1 \end{pmatrix}$

(c) $\mathbf{A}^2 = \begin{pmatrix} 4+15 & -6-12 \\ -10-20 & 15+16 \end{pmatrix} = \begin{pmatrix} 19 & -18 \\ -30 & 31 \end{pmatrix}$

(d) $\mathbf{B}^2 = \begin{pmatrix} 1+18 & -6+12 \\ -3+6 & 18+4 \end{pmatrix} = \begin{pmatrix} 19 & 6 \\ 3 & 22 \end{pmatrix}$

6. (a) $\mathbf{AB} = \begin{pmatrix} 5 & -6 & 7 \end{pmatrix}\begin{pmatrix} 3 \\ 4 \\ -1 \end{pmatrix} = (-16)$

(b) $\mathbf{BA} = \begin{pmatrix} 3 \\ 4 \\ -1 \end{pmatrix} (5 \quad -6 \quad 7) = \begin{pmatrix} 15 & -18 & 21 \\ 20 & -24 & 28 \\ -5 & 6 & -7 \end{pmatrix}$

(c) $(\mathbf{BA})\mathbf{C} = \begin{pmatrix} 15 & -18 & 21 \\ 20 & -24 & 28 \\ -5 & 6 & -7 \end{pmatrix} \begin{pmatrix} 1 & 2 & 4 \\ 0 & 1 & -1 \\ 3 & 2 & 1 \end{pmatrix} = \begin{pmatrix} 78 & 54 & 99 \\ 104 & 72 & 132 \\ -26 & -18 & -33 \end{pmatrix}$

(d) Since $\mathbf{AB}$ is 1×1 and $\mathbf{C}$ is 3×3 the product $(\mathbf{AB})\mathbf{C}$ is not defined.

9. (a) $(\mathbf{AB})^T = \begin{pmatrix} 7 & 10 \\ 38 & 75 \end{pmatrix}^T = \begin{pmatrix} 7 & 38 \\ 10 & 75 \end{pmatrix}$

(b) $\mathbf{B}^T\mathbf{A}^T = \begin{pmatrix} 5 & -2 \\ 10 & -5 \end{pmatrix} \begin{pmatrix} 3 & 8 \\ 4 & 1 \end{pmatrix} = \begin{pmatrix} 7 & 38 \\ 10 & 75 \end{pmatrix}$

12. $\begin{pmatrix} 6t \\ 3t^2 \\ -3t \end{pmatrix} + \begin{pmatrix} -t+1 \\ -t^2+t \\ 3t-3 \end{pmatrix} - \begin{pmatrix} 6t \\ 8 \\ -10t \end{pmatrix} = \begin{pmatrix} -t+1 \\ 2t^2+t-8 \\ 10t-3 \end{pmatrix}$

15. Since $\det \mathbf{A} = 0$, $\mathbf{A}$ is singular.

18. Since $\det \mathbf{A} = -6$, $\mathbf{A}$ is nonsingular.

$$\mathbf{A}^{-1} = -\frac{1}{6}\begin{pmatrix} 2 & -10 \\ -2 & 7 \end{pmatrix}$$

21. Since $\det \mathbf{A} = -9$, $\mathbf{A}$ is nonsingular. The cofactors are

$$\begin{array}{lll} A_{11} = -2 & A_{12} = 13 & A_{13} = 8 \\ A_{21} = -2 & A_{22} = 5 & A_{23} = -1 \\ A_{31} = -1 & A_{32} = 7 & A_{33} = -5. \end{array}$$

Then

$$\mathbf{A}^{-1} = -\frac{1}{9}\begin{pmatrix} -2 & -13 & 8 \\ -2 & 5 & -1 \\ -1 & 7 & -5 \end{pmatrix}^T = -\frac{1}{9}\begin{pmatrix} -2 & -2 & -1 \\ -13 & 5 & 7 \\ 8 & -1 & -5 \end{pmatrix}.$$

24. Since $\det \mathbf{A}(t) = 2e^{2t} \neq 0$, $\mathbf{A}$ is nonsingular.

$$\mathbf{A}^{-1} = \frac{1}{2}e^{-2t}\begin{pmatrix} e^t \sin t & 2e^t \cos t \\ -e^t \cos t & 2e^t \sin t \end{pmatrix}$$

27. $X = \begin{pmatrix} 2e^{2t} + 8e^{-3t} \\ -2e^{2t} + 4e^{-3t} \end{pmatrix}$ so that $\dfrac{dX}{dt} = \begin{pmatrix} 4e^{2t} - 24e^{-3t} \\ -4e^{2t} - 12e^{-3t} \end{pmatrix}$.

30. (a) $\dfrac{dA}{dt} = \begin{pmatrix} -2t/(t^2+1)^2 & 3 \\ 2t & 1 \end{pmatrix}$

(b) $\dfrac{dB}{dt} = \begin{pmatrix} 6 & 0 \\ -1/t^2 & 4 \end{pmatrix}$

(c) $\displaystyle\int_0^1 A(t)\,dt = \begin{pmatrix} \tan^{-1} t & \frac{3}{2}t^2 \\ \frac{1}{3}t^3 & \frac{1}{2}t^2 \end{pmatrix}\Big|_{t=0}^{t=1} = \begin{pmatrix} \frac{\pi}{4} & \frac{3}{2} \\ \frac{1}{3} & \frac{1}{2} \end{pmatrix}$

(d) $\displaystyle\int_1^2 B(t)\,dt = \begin{pmatrix} 3t^2 & 2t \\ \ln t & 2t^2 \end{pmatrix}\Big|_{t=1}^{t=2} = \begin{pmatrix} 9 & 2 \\ \ln 2 & 6 \end{pmatrix}$

(e) $A(t)B(t) = \begin{pmatrix} 6t/(t^2+1) + 3 & 2/(t^2+1) + 12t^2 \\ 6t^3 + 1 & 2t^2 + 4t^2 \end{pmatrix}$

(f) $\dfrac{d}{dt}A(t)B(t) = \begin{pmatrix} (6 - 6t^2)/(t^2+1)^2 & -4t/(t^2+1)^2 + 24t \\ 18t^2 & 12t \end{pmatrix}$

(g) $\displaystyle\int_1^t A(s)B(s)\,ds = \begin{pmatrix} 6s/(s^2+1) + 3 & 2/(s^2+1) + 12s^2 \\ 6s^3 + 1 & 6s^2 \end{pmatrix}\Big|_{s=1}^{s=t}$

$= \begin{pmatrix} 3t + 3\ln(t^2+1) - 3 - 3\ln 2 & 4t^3 + 2\tan^{-1} t - 4 - \pi/2 \\ (3/2)t^4 + t - (5/2) & 2t^3 - 2 \end{pmatrix}$

33. $\begin{pmatrix} 1 & -1 & -5 & | & 7 \\ 5 & 4 & -16 & | & -10 \\ 0 & 1 & 1 & | & -5 \end{pmatrix} \Longrightarrow \begin{pmatrix} 1 & -1 & -5 & | & 7 \\ 0 & 1 & 1 & | & -5 \\ 0 & 9 & 9 & | & -45 \end{pmatrix} \Longrightarrow \begin{pmatrix} 1 & 0 & -4 & | & 2 \\ 0 & 1 & 1 & | & -5 \\ 0 & 0 & 0 & | & 0 \end{pmatrix}$

Letting $z = t$ we find $y = -5 - t$, and $x = 2 + 4t$.

36. $\begin{pmatrix} 1 & 0 & 2 & | & 8 \\ 1 & 2 & -2 & | & 4 \\ 2 & 5 & -6 & | & 6 \end{pmatrix} \Longrightarrow \begin{pmatrix} 1 & 0 & 2 & | & 8 \\ 0 & 2 & -4 & | & -4 \\ 0 & 5 & -10 & | & -10 \end{pmatrix} \Longrightarrow \begin{pmatrix} 1 & 0 & 2 & | & 8 \\ 0 & 1 & -2 & | & -2 \\ 0 & 0 & 0 & | & 0 \end{pmatrix}$

Letting $z = t$ we find $y = -2 + 2t$, and $x = 8 - 2t$.

39.
$$\begin{pmatrix} 1 & 2 & 4 & | & 2 \\ 2 & 4 & 3 & | & 1 \\ 1 & 2 & -1 & | & 7 \end{pmatrix} \implies \begin{pmatrix} 1 & 2 & 4 & | & 2 \\ 0 & 0 & -5 & | & -3 \\ 0 & 0 & -5 & | & 5 \end{pmatrix} \implies \begin{pmatrix} 1 & 2 & 0 & | & -2/5 \\ 0 & 0 & 1 & | & 3/5 \\ 0 & 0 & 0 & | & 8 \end{pmatrix}$$

There is no solution.

42. We solve

$$\det(\mathbf{A} - \lambda\mathbf{I}) = \begin{vmatrix} 2 - \lambda & 1 \\ 2 & 1 - \lambda \end{vmatrix} = \lambda(\lambda - 3) = 0.$$

For $\lambda_1 = 0$ we have

$$\begin{pmatrix} 2 & 1 & | & 0 \\ 2 & 1 & | & 0 \end{pmatrix} \implies \begin{pmatrix} 1 & 1/2 & | & 0 \\ 0 & 0 & | & 0 \end{pmatrix}$$

so that $k_1 = -\frac{1}{2}k_2$. If $k_2 = 2$ then

$$\mathbf{K}_1 = \begin{pmatrix} -1 \\ 2 \end{pmatrix}.$$

For $\lambda_2 = 3$ we have

$$\begin{pmatrix} -1 & 1 & | & 0 \\ 2 & -2 & | & 0 \end{pmatrix} \implies \begin{pmatrix} 1 & -1 & | & 0 \\ 0 & 0 & | & 0 \end{pmatrix}$$

so that $k_1 = k_2$. If $k_2 = 1$ then

$$\mathbf{K}_2 = \begin{pmatrix} 1 \\ 1 \end{pmatrix}.$$

45. We solve

$$\det(\mathbf{A} - \lambda\mathbf{I}) = \begin{vmatrix} 5 - \lambda & -1 & 0 \\ 0 & -5 - \lambda & 9 \\ 5 & -1 & -\lambda \end{vmatrix} = \begin{vmatrix} 4 - \lambda & -1 & 0 \\ 4 - \lambda & -5 - \lambda & 9 \\ 4 - \lambda & -1 & -\lambda \end{vmatrix} = \lambda(4 - \lambda)(\lambda + 4) = 0.$$

If $\lambda_1 = 0$ then

$$\begin{pmatrix} 5 & -1 & 0 & | & 0 \\ 0 & -5 & 9 & | & 0 \\ 5 & -1 & 0 & | & 0 \end{pmatrix} \implies \begin{pmatrix} 1 & 0 & -9/25 & | & 0 \\ 0 & 1 & -9/5 & | & 0 \\ 0 & 0 & 0 & | & 0 \end{pmatrix}$$

so that $k_1 = \frac{9}{25}k_3$ and $k_2 = \frac{9}{5}k_3$. If $k_3 = 25$ then

$$\mathbf{K}_1 = \begin{pmatrix} 9 \\ 45 \\ 25 \end{pmatrix}.$$

If $\lambda_2 = 4$ then

$$\begin{pmatrix} 1 & -1 & 0 & | & 0 \\ 0 & -9 & 9 & | & 0 \\ 5 & -1 & -4 & | & 0 \end{pmatrix} \implies \begin{pmatrix} 1 & 0 & -1 & | & 0 \\ 0 & 1 & -1 & | & 0 \\ 0 & 0 & 0 & | & 0 \end{pmatrix}$$

so that $k_1 = k_3$ and $k_2 = k_3$. If $k_3 = 1$ then

$$\mathbf{K}_2 = \begin{pmatrix} 1 \\ 1 \\ 1 \end{pmatrix}.$$

If $\lambda_3 = -4$ then

$$\begin{pmatrix} 9 & -1 & 0 & | & 0 \\ 0 & -1 & 9 & | & 0 \\ 5 & -1 & 4 & | & 0 \end{pmatrix} \implies \begin{pmatrix} 1 & 0 & -1 & | & 0 \\ 0 & 1 & -9 & | & 0 \\ 0 & 0 & 0 & | & 0 \end{pmatrix}$$

so that $k_1 = k_3$ and $k_2 = 9k_3$. If $k_3 = 1$ then

$$\mathbf{K}_3 = \begin{pmatrix} 1 \\ 9 \\ 1 \end{pmatrix}.$$

48. We solve

$$\det(\mathbf{A} - \lambda\mathbf{I}) = \begin{vmatrix} 1 - \lambda & 6 & 0 \\ 0 & 2 - \lambda & 1 \\ 0 & 1 & 2 - \lambda \end{vmatrix} = \begin{vmatrix} 1 - \lambda & 6 & 0 \\ 0 & 3 - \lambda & 3 - \lambda \\ 0 & 1 & 2 - \lambda \end{vmatrix} = (3 - \lambda)(1 - \lambda)^2 = 0.$$

For $\lambda = 3$ we have

$$\begin{pmatrix} -2 & 6 & 0 & | & 0 \\ 0 & 0 & 0 & | & 0 \\ 0 & 1 & -1 & | & 0 \end{pmatrix} \implies \begin{pmatrix} 1 & 0 & -3 & | & 0 \\ 0 & 1 & -1 & | & 0 \\ 0 & 0 & 0 & | & 0 \end{pmatrix}$$

so that $k_1 = 3k_3$ and $k_2 = k_3$. If $k_3 = 1$ then

$$\mathbf{K}_1 = \begin{pmatrix} 3 \\ 1 \\ 1 \end{pmatrix}.$$

For $\lambda_2 = \lambda_3 = 1$ we have

$$\begin{pmatrix} 0 & 6 & 0 & | & 0 \\ 0 & 1 & 1 & | & 0 \\ 0 & 1 & 1 & | & 0 \end{pmatrix} \implies \begin{pmatrix} 0 & 1 & 0 & | & 0 \\ 0 & 0 & 1 & | & 0 \\ 0 & 0 & 0 & | & 0 \end{pmatrix}$$

so that $k_2 = 0$ and $k_3 = 0$. If $k_1 = 1$ then

$$\mathbf{K}_2 = \begin{pmatrix} 1 \\ 0 \\ 0 \end{pmatrix}.$$

51. Let

$$\mathbf{A} = \begin{pmatrix} a_{11} & a_{12} \\ a_{21} & a_{22} \end{pmatrix}.$$

Then

$$\frac{d}{dt}[\mathbf{A}(t)\mathbf{X}(t)] = \frac{d}{dt}\begin{pmatrix} a_1 & a_2 \\ a_3 & a_4 \end{pmatrix}\begin{pmatrix} x_1 \\ x_2 \end{pmatrix} = \frac{d}{dt}\begin{pmatrix} a_1 x_1 + a_2 x_2 \\ a_3 x_1 + a_4 x_2 \end{pmatrix} = \begin{pmatrix} a_1 x_1' + a_1' x_1 + a_2 x_2' + a_2' x_2 \\ a_3 x_1' + a_3' x_1 + a_4 x_2' + a_4' x_2 \end{pmatrix}$$

$$= \begin{pmatrix} a_1 & a_2 \\ a_3 & a_4 \end{pmatrix}\begin{pmatrix} x_1' \\ x_2' \end{pmatrix} + \begin{pmatrix} a_1' & a_2' \\ a_3' & a_4' \end{pmatrix}\begin{pmatrix} x_1 \\ x_2 \end{pmatrix} = \mathbf{A}(t)\mathbf{X}'(t) + \mathbf{A}'(t)\mathbf{X}(t).$$

54. Since

$$(\mathbf{AB})(\mathbf{B}^{-1}\mathbf{A}^{-1}) = \mathbf{A}(\mathbf{BB}^{-1})\mathbf{A}^{-1} = \mathbf{AIA}^{-1} = \mathbf{AA}^{-1} = \mathbf{I}$$

and

$$(\mathbf{B}^{-1}\mathbf{A}^{-1})(\mathbf{AB}) = \mathbf{B}^{-1}(\mathbf{A}^{-1}\mathbf{A})\mathbf{B} = \mathbf{B}^{-1}\mathbf{IB} = \mathbf{B}^{-1}\mathbf{B} = \mathbf{I}$$

we have

$$(\mathbf{AB})^{-1} = \mathbf{B}^{-1}\mathbf{A}^{-1}.$$